THE AUTHORS

Oleg Polunin was born in Reading, Berkshire, in 1914 of
Russian/English parentage. After reading botany at Oxford
University, he started teaching at Charterhouse school in 1938.
After the war, he was invited to join a climbing expedition to
Nepal, and this led to other botanical trips, to Turkey, the
Middle East, Kashmir, Karakoram and Greece, the last inspir-
ing him and Anthony Huxley to write *Flowers of the Mediterra-
nean* (1965) of which this is an updated edition. More books,
based on his continuing travels, followed, such as *Flowers of
Europe* (1985), *Flowers of the Himalayas* (1984), and *Guide to
the Vegetation of Europe* (1969). A skilled photographer, Oleg
Polunin illustrated his own books. In 1984 he received the
Bloomer Award from the Linnaean Society of London for his
work in botany. He was married with two children, and died in
1985.

Anthony Huxley, who was born in 1920, the elder son of Sir
Julian Huxley, F.R.S., is well known as a botanical and
horticultural authority. He was for many years on the staff,
and finally editor, of *Amateur Gardening* magazine, and now
works as a freelance. To date, he has written or edited 33
books. These include several field guides, the wide-ranging
Plant and Planet (1974), *An Illustrated History of Gardening*
(1978), *The Penguin Encyclopaedia of Gardening* (1981), *The
Macmillan World Guide to House Plants* (1983), and the latest,
Green Inheritance (1984), written for the World Wildlife Fund.
A consuming interest in plant life has taken him on travels all
over the world. He is a member of the Council of the Royal
Horticultural Society, which has awarded him its highest
accolade, the Victoria Medal of Honour.

ACKNOWLEDGMENTS

Many people have helped in the preparation of this book. We are in the first place indebted to the Director, Royal Botanic Gardens, Kew, and the Regius Keeper, Royal Botanic Garden, Edinburgh, for the initial identification of our field collections, and to the latter for the loan of further herbarium specimens. Further identifications were made by the Department of Botany, British Museum (Natural History).

Individuals to whom our special gratitude is due include the following:

Mr R.D. Meikle, Mr C. C. Townsend, Dr Peter Davis and Mr Ian Hedge for advice, and help with identification problems.

Dr V. H. Heywood, Dr S.M. Walters and the late Dr J. S. L. Gilmour for advice on nomenclature.

Mr Richard Gorer for reading the manuscript and making many pertinent suggestions.

Mr Roy Hay for the loan of rare Floras, and Mr P. Stageman (lately Librarian of the Royal Horticultural Society) for his never-failing help.

Mrs Barbara Everard for her enthusiastic co-operation in making the drawings, largely from herbarium specimens.

Most of the photographs were taken by us, but we thank three people for filling gaps - Mr M. D'Oyly for illustrations nos. 29, 39, 44, 46, 128, 135, 213, 251 and 270; Miss Daphne Barry for no. 11; and Mr C. C. Townsend for no. 111.

Lastly but by no means least we thank the Society for Hellenic Travel, without whose agency we could not have covered the ground for our studies.

NOTE

In this third edition of *Flowers in the Mediterranean* the opportunity has been taken to update the nomenclature by preparing a list of alterations recommended in *Flora Europaea* and other relevant recent authorities; to add newer works to the Bibliography; and to make some very small textual changes.

Alas, I have not been able to share this work with my co-author Oleg Polunin, who died in 1985. That this book has remained in print for over 20 years is very much a measure of the care he put into our joint venture so long ago.

Anthony Huxley

FLOWERS OF THE MEDITERRANEAN

Oleg Polunin
and Anthony Huxley

*With 311 illustrations in colour
from photographs by the Authors and
128 line drawings by Barbara Everard*

THE HOGARTH PRESS

LONDON

To Eileen Hardie and Anne Hindle,
without whom this book could not
have been written

Published in 1987 by
The Hogarth Press
Chatto and Windus Ltd
30 Bedford Square, London WC1B 3RP

First published in Great Britain by Chatto and Windus Ltd 1965
Hogarth edition offset, with additions and revisions, from original
British edition
Copyright © Anthony Huxley and the Estate of Oleg Polunin
1965, 1972, 1987

British Library Cataloguing in Publication Data

Polunin, Oleg
Flowers of the Mediterranean.–New, updated ed.
1.Wild flowers–Mediterranean Region
I. Title II. Huxley,Anthony
582.13'09182'2   QK314.5

ISBN 0-7012-0784-1

Printed in Great Britain by
Redwood Burn Ltd
Trowbridge, Wilts

CONTENTS

Illustration References. The bold numerals in the descriptive text (pages 51 to 243) refer to the illustrations. These are numbered consecutively, numbers 1 to 311 being in colour and numbers 312 to 439 being line drawings immediately following the colour plates.

To locate the text pages upon which illustrated plants are described, it is necessary to refer to the plant's name in the Index, where both text pages and figure references are printed.

INTRODUCTION

THE PURPOSE OF THIS BOOK

The Mediterranean flora is exceptionally rich and colourful, and more and more people are visiting the lands of the Mediterranean, particularly those places where the ruins of ancient civilizations are to be seen – Sicily, Italy, Greece and its islands, Turkey, the Lebanon – and of course the Holy Land, full of its reminders of early Christian times. For many years there has been no book in print to assist the tourist in naming the flowers he is likely to see; the few popular books are long out of print, and the authoritative floras, where they exist at all, are either out of print, extremely expensive, in Latin, or unillustrated – if not all these at once!

This book is designed to fill this need for a well-illustrated, straightforward guide to the commoner flowers of the Mediterranean. The plants described have been selected to include the most beautiful and interesting, as well as the most abundant species, which are likely to be seen by the visitor to Mediterranean countries, especially near the coasts, but also on excursions inland. The whole Mediterranean is covered, from Morocco and Spain in the west, the Riviera, Italy, the Adriatic, Greece and the Aegean, to the Turkish coast, Lebanon, Syria and Israel in the east. A number of the cultivated and naturalized plants are also included since they form an important part of the botanical scene.

Over 700 species have been described, in terms as simple and non-technical as possible, though it is hoped in sufficient detail to be of value to the serious botanist. In each description an attempt has been made to give a general picture of the species, before the more detailed description which follows. Our aim is not only to enable the reader to spot and name the plant, but to be able to satisfy himself with reasonable certainty, by checking through the description with the plant itself, that he has in fact made the right identification. In addition to the description, which includes habitat, distribution, and time of flowering, reference is made to the uses and importance of the plants in antiquity, in the Bible, and at the present day.

311 plants are illustrated from colour photographs taken in their natural surroundings, and a further 128 plants are shown in line drawings. Botanical details are shown in a further 61 drawings in the text.

OUR SELECTION OF SPECIES

One of the most difficult tasks facing us at the start was to decide which species to include in a popular handbook of Mediterranean flowers, inevitably restricted in size. We felt strongly that the approach to the plants had to be primarily pictorial, and not a mere series of descriptions which the non-botanical reader would find difficult to visualize. Our choice, in the first place, has therefore been influenced by the photographic material amassed in the course of numerous visits to different parts of the area, and the lists of species noted during these visits. Popular floras were also studied. Experience in conducting botanical parties gave us some idea of the plants which habitually arouse curiosity, and also those which do not!

Of the large numbers of species that could be included in a list of Mediterranean plants, many have been excluded because they have insignificant flowers and are thus of less interest to the layman. Others occur too sporadically for inclusion, in the areas on which this book concentrates – the littoral, lower hills and the few higher areas such as Delphi in Greece where tourists are most likely to go.

The total number of flowering plants found in the Mediterranean regions is large but difficult to assess. Greece alone is estimated to have over 6,000; Mount Parnes outside Athens has 1,000; there are 1,000 within five miles radius of Jerusalem, and a tiny area like the Department of Var in the French Riviera has over 2,000 (more, incidentally, than all Great Britain). Many species occur throughout the region, but many more are local or endemic plants restricted to certain countries, islands or mountain tops.

Almost every genus of plants mentioned in the text is represented by at least one photograph or drawing; closely allied species are described in less detail, and we hope that the illustrations will enable other related species to be recognized as such by their family likeness. Some of the more extensive groups have inevitably had to be represented by a mere one or two members. A special feature has, however, been made of the Orchid family, which always arouses a good deal of interest but is difficult to master without good pictorial references.

It will be noticed that the order of the colour illustrations does not always follow that of the text. This is simply because the jigsaw puzzle of laying out illustrations of different shapes and sizes sometimes imposed itself above the strict botanical order. The colour illustrations are, however, in family order, and every effort has been made to keep similar and closely related plants together.

In the black and white plates reasons of expediency made it necessary to depart on occasion from the family order.

The colour and line illustrations naturally overlap each other, but for ease of reference they are numbered consecutively, the line illustrations following on after the colour.

BOTANICAL INFORMATION

Wherever possible plants have been described from actual specimens collected by the authors or herbarium material loaned from other collections; the drawings have likewise been made largely from these specimens. The always present problem of achieving acceptable botanical accuracy has been somewhat complicated. When this book was written, the first volume of *Flora Europaea* had appeared, and it was obviously desirable to follow the treatment of this unifying work, produced in consultation with leading European botanists. Unfortunately, later volumes altered many of the names regarded as correct at the time of writing. A summary of all changes in botanical names recommended in *Flora Europaea* has been supplied on pp.ix–xi for this third edition of *Flowers of the Mediterranean*. The ideal solution of altering the text rather than supplying this list was not feasible owing to the heavy expense of resetting, especially with the colour plate captions.

Of course taxonomic research and name alteration continue all the time. But basically no attempt has been made to follow up such work after the publication of *Flora Europaea* (1964–1980), which has to be considered as a botanical consensus for the next decade – at least by non-experts.

The few plants which exist only in Turkey, Syria, Lebanon and Israel have been checked in appropriate sources, notably the recently completed *Flora of Turkey*.

The order of families follows *Flora Europaea* and uses the Englerian order so widely used on the Continent, with the exception that the mono-cotyledons are placed at the end. This order differs fundamentally from that commonly used in British Floras based on the system of Bentham & Hooker.

It was not, however, possible to place the genera and species in the order finally used in *Flora Europaea*, for at the time of writing this had not been finally settled; and in consequence we have largely followed the arrangement in K. H. Rechinger's *Flora Aegaea*, with appropriate interpolations for plants outside the area covered. The arrangement of *Orchidaceae* is modelled on that of Clapham, Tutin and Warburg in *Flora of the British Isles*, while the detail of the genus *Ophrys* follows that of Erich Nelson (op. cit. p. 244), subject to recent taxonomic changes.

In some cases two or more Latin names are given for a single plant: the names in brackets are the synonyms, commonly used names which are likely to be found in older floras but which research has now shown to be invalid.

The names or abbreviations after each Latin name are those of the botanists responsible for describing the species in the first place. Such definition of the authority is essential particularly when it happens that the same specific (second) name is applied to more than one plant.

English names for Mediterranean plants are liable to be suspect and the only criterion of their veracity is that of common usage. Care has been taken not to 'coin' any English names, and they have only been used when it is considered that they are widely known.

The metric system has been used for all measurements of dimensions in accordance with modern scientific practice.

CHANGES IN BOTANICAL NAMES

Reference is made on pp. vii–viii to nomenclatural problems, and the desirability of following the nomenclature adopted in *Flora Europaea*. The list below embodies changes made in that work, which are not in the existing text.

Name in book	Recommended new name	Page
Acanthus spinosissimus	**A.spinosus L.**	172
Aegilops ovata	**A.geniculata** Roth. and **A.neglecta** Req. ex Bertol.	198
Ajuga chia	**A.chamaepitys** (L.)Schreber ssp. **chia** (Schreber)Arcangeli	156
Alkanna boeotica	**A.graeca** Boiss. & Sprun.ssp. **chia** (DC.)Nyman	151
Allium sphaerocephalum	now spelt **A.sphaerocephalon**	210
Althaea cretica	**Alcea pallida** (Willd.)Waldst. & Kit. ssp.**cretica**(Weinm.)D.A.Webb	125
Anagallis arvensis ssp. foemina	**A.foemina** Mill.	142
Anchusa hybrida	**A.undulata** L.	150
Andropogon hirtus	**Hyparrhenia hirta** (L.)Stapf	203
Anemone hortensis var.pavonina	**A.pavonina** Lam.	70
Anthyllis vulneraria ssp.spruneri	**A.v.**ssp.**praepropera** (A.Kerner) Bornm.	100
Antirrhinum orontium	**Misopates orontium** (L.)Rafin.	168
Aspodelus microcarpus	**A.aestivus** Brot.	208
Bellis silvestris	now spelt **B.sylvestris**	183
Bonjeania hirsuta	**Dorycnium hirsutum** (L.)Ser.	102
Calycotome infesta	**Calicotome spinosa** (L.)Link	93
Campanula spathulata	now spelt **C.spatulata**	179
Catapodium rigidum	**Dezmazaria rigida** Tutin	200
Centranthus calcitrapa	now spelt **C.calcitrapae**	176
Cichorium pumilum	**C.endivia**L.ssp.**divaricatum** (Shousboe) P.D.Sell	193
Cistus villosus	**C.incanus** L.	128
Cistus villosus ssp. creticus	**C.incanus** L.ssp.**creticus**(L.) Heywood	128
Cistus ladaniferus	now spelt **C.ladanifer**	129
Cistus salviaefolius	now spelt **C.salvifolius**	129
Citrus maxima	**C.paradisi** Macfadyen	114
Citrus nobilis	**C.deliciosa** Ten.	114
Convolvulus cantabricum	now spelt **C.cantabrica**	147
Convolvulus elegantissimus	**C.althaeoides** L.ssp.**tenuissimus** (S. & S.)Stace	148
Coridothymus capitatus	**Thymus capitatus** (L.)Hoffmans. & Link	163
Crocus pallasii var. cartwrightianus	**C.cartwrightianus** Herb. (*C.pallasii* is a distinct, more easterly sp.)	223
Cyclamen neapolitanum	**C.hederifolium** Aiton	141
Cyclamen orbiculatum	**C.vernum** Sw.	142
Cytisus triflorus	**C.villosus** Poiret	93
Dactylorchis romana	**Dactylorhiza sulphurea** (Ten) Franco ssp.**pseudosambucina** (Seb. & Maur.)Soó	241
Datura metel	Plants found in the Mediterranean are actually **D.innoxia** Miller	166
Echinops viscosus	**E.spinosissimus** Turra	153
Echium diffusum	**E.angustifolium** Miller	153
Echium lycopsis	**E.plantagineum** L.	154

Fig. 1. Mean February temperatures in the Mediterranean. Figures are degrees Fahrenheit.

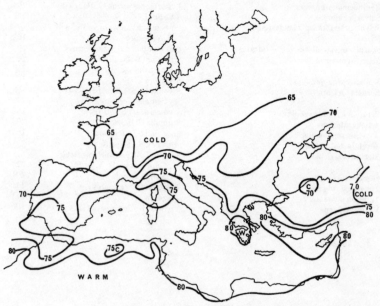

Fig. 2. Mean August temperatures in the Mediterranean. Figures are degrees Fahrenheit.

THE VEGETATION
OF THE MEDITERRANEAN

The variety and richness of plant life found in the Mediterranean basin is proverbial. Man made some of his earliest settlements here, and by domesticating the wild fruits of the region such as the grape, olive, fig, carob, and cereal grasses, he found an abundance which enabled him to develop his skills and establish civilizations.

For at least 8,000 years he has been cutting wood for building, carving, fuel, smelting and tanning; growing crops and cultivating orchards; and grazing his sheep and goats; with the result that today there is little left of the natural covering of vegetation. All that now remains between the cultivated valleys is a patchwork of plant life in all stages of regeneration and degeneration, for we must think of plant communities like heaths and thickets as short-lived, slowly changing stages in a gradual trend towards a stable balance between vegetation, climate and soil. The forest is, in almost all cases, the stable endpoint, or climatic-climax as the ecologists call it, yet how infrequently do we see mature forests developed in the Mediterranean region.

So continuously and over such long periods have goats nibbled, trees been felled and brushwood cleared that the soil has been lost and the hillsides have become open rocky slopes with a few scattered bushes and grey-leaved herbs – now such a familiar condition of many thousands of square miles of the Mediterranean coastline.

CLIMATE

The Mediterranean climate is so extremely varied that no satisfactory single criterion has been found to define it. The overall picture is clear enough. Rain falls during the months of October to April, but the winters are not cold and average over 40°F (4·3°C) to above 50°F (10°C) over much of the area during the coldest months (see Fig. 1). Frosts are infrequent. The summers are very hot with mean temperatures over 70°F (21°C) and rising to 86°F (30°C) in North Africa (see Fig. 2). The summer days are very bright and sunny and this is the period of 'la grande chaleur' when the average sunshine is over 10 hours per day during the months of June to August. This hot season is ideal for ripening fruits of all kinds for which the Mediterranean region is justly famous.

Most plant growth ceases during this hot period and only begins again when the first rains arrive, usually in October (see Fig. 3). Some species flower during late autumn and early winter, and some never stop active growth at all during the winter. Early spring is the time when most Mediterranean perennials flower, and the flowering period rises to its peak towards the end of April when in addition a rich variety of annual plants carpet the plains and hillsides. By June they have died down and many have shed their seeds, and only the thistles and members of the Mint family (*Labiatae*) are likely to be found in flower.

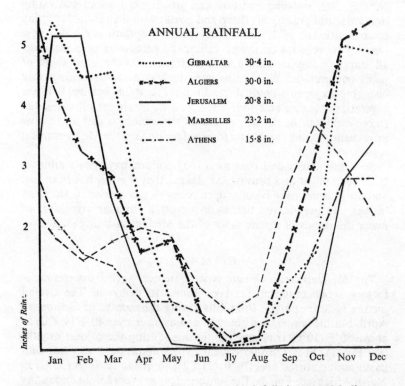

Fig. 3. Average monthly rainfall, and total rainfall, in some Mediterranean cities.

2

Emberger has attempted to give an empirical formula for the Mediterranean climate in the form $\frac{R}{M}$ where $R = $ total mm. of rain in the months June, July, August, and $M = $ mean maximum temperature (°C) of the hottest month. Regions in the Mediterranean basin with a quotient below 7 are said to have a true Mediterranean climate.

Many people have thought of the Olive as the plant indicator of the Mediterranean climate for it will only flourish in a typically Mediterranean environment; it will not tolerate an average temperature below 37·4°F (3°C) during the coldest winter month – it will grow to altitudes of 600–800 m. (see Fig. 4).

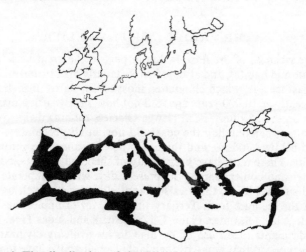

Fig. 4. The distribution of the Olive in the Mediterranean basin.

However, the main objections to using the Olive as an indicator are that it is probably not a native of the Mediterranean, but of more eastern origin, and that it is a cultivated plant. Others have suggested the Holm Oak as a truer wild indicator plant, but it has a markedly western distribution. However, wherever one sees the Olive, Holm Oak, the Kermes Oak or the Aleppo Pine growing, and particularly where any two of these are growing together, one can be pretty certain that one is in a Mediterranean climate. There are, of course, many other plants that only flourish in this climate and these we call Mediterranean species, e.g. Flahault records 705 Mediterranean

3

species in the French Flora; Turrill 700 species in the Balkan Peninsula.

The overall pattern of the Mediterranean climate is profoundly affected by the surrounding regions, and in general moist air comes from the Atlantic, cold air from central Europe and Turkey, and hot air from the Sahara. Thus western coasts usually have more rainfall than the corresponding eastern coasts; winds are many and varied, and are of great importance locally to plant life. The cool 'mistral' and 'bora' blowing down from central Europe effectively limits the extension of the Mediterranean plants; while the hot 'sirocco' blowing from the Sahara makes a semi-desert of much of the North African coastline, or again the summer 'etesian' winds of Greece or 'meltemi' of the Turks bring down hot air from Central Russia.

THE VARIETY OF PLANT LIFE

The richness of the flora is due in part to the great variations in climate and habitat, and also no less to its early history and evolution. The last ice age, which eliminated most of our native British species as recently as 10,000 years ago, did not have nearly such a disastrous effect on the Mediterranean plants. Glaciers running down from the mountains surrounding the coast did not, in all probability, extend much below 2,700 m., and there was no permanent ice sheet in the vicinity. There must have been plenty of sheltered pockets, no doubt moister and cooler than at the present day, where temperate plants could survive. Thus, there exist many 'old' species which have survived the ice ages from Tertiary times as, for example, the Carob, Myrtle, Vine, Oleander, Plane, Olive, Lentisk and Judas Tree – which include plants made use of by man in his evolving civilization. It will be noticed that many of the plants named above have few or no close relatives, indicating that they have diverged from the main stock a long time ago, and all closely related species have become extinct. The Carob is the sole survivor of some ancient stock; it is the only species in the genus *Ceratonia*.

Recently evolved species and those that are now actively evolving new species and sub-species are often easily recognized by the very fact that they are difficult to distinguish from each other. They hybridize, they show graduations from one to another, and have local variants which are often a cause of disagreement by botanists and a testing ground for botanical terminology, to the despair of the amateur botanist. The Mulleins, Campanulas, Centaureas, *Ophrys*, Daisy family and Pea family are examples of rapidly evolving genera

4

and families, and cause more trouble to identify than most other groups.

A study of plants with local distributions, that is to say 'endemic' species, is very revealing. The Mediterranean flora is rich in endemics. The flooding of the Mediterranean something like a million years ago, and the consequent dissection of the land into islands and peninsulas, has created barriers to breeding and interchange. Species isolated on islands or mountain tops have developed along different lines from their relatives on the mainland; separated thus for thousands of years they have become quite distinct and are given different names. Greece is perhaps the most notable Mediterranean country in this respect, with about one plant in five being an endemic; the small Athos peninsula has 16 and Mount Olympus 19 endemics.

Turrill, in *Plant Life of the Balkan Peninsula*, has estimated that one in four of all Balkan plants are endemic to the region, of which about 1 per cent are ancient species, 16 per cent are old or fairly old, and 70 per cent are fairly recent or recently evolved, the remainder being unknown. But plant populations are not static; migrations have occurred and are occurring today. The Balkans were once in the Tertiary period a peninsula of Turkey so that they are now rich in the Irano-Turanian plants of Asia Minor and Persia. To a lesser extent plants have come from the Pontic region round the Black Sea, and from the Eastern Alps of Europe, but not from North Africa.

Thus the continuous and possibly increasing rate of evolutionary advance, the geographical isolation, the migrations of plants and local climatic variations have all contributed towards making Mediterranean plant life as rich and varied as in any part of the world. Added to this, man has, during the last few hundred years, brought in many sub-tropical plants from all over the world to grow in his gardens and orchards, and to plant along his boulevards. The Palms, Loquat, Agaves, Cacti, Mimosas, Eucalyptus, Oranges and Lemons are all foreigners, yet how familiar they have become in the Mediterranean landscape; they now seem to be an integral part of it.

PLANT COMMUNITIES

Wild plants colonizing a tract of land will slowly develop into a plant community of closely inter-reacting, and later inter-dependent species. Soil will gradually develop and deepen; animals will come to play an important part in the community. Stage by stage, over the years, a successive series of developmental communities, or 'seres' as they are called technically, will lead to a more or less stable plant community which we call a climax. The climax plant community which can develop naturally in a Mediterranean climate is a forest of evergreen trees with leathery leaves, commonly dominated either by evergreen Oaks or Pines. These trees shelter a number of subsidiary layers of vegetation; usually a shrub layer of small-leaved evergreen shrubs, and a field layer of herbaceous perennial plants. This 'climatic-climax' vegetation may remain relatively stable for centuries if undisturbed by man, but today all we are likely to find of this 'primaeval' forest are small patches tucked away in remote and inaccessible valleys in the hills.

Local variations of soil, climate, exposure and altitude may prevent this climatic-climax from developing, and a stable community, often of lower status, such as maquis, may result.

Although these climax communities can remain unchanged for long periods, once the forest trees are cleared a rapid degeneration is likely to follow. Further cutting of brushwood and grazing by animals can, in relatively few years, result in the complete degradation of the forest into a steppe-like landscape with a few grey-leaved shrubs and herbaceous plants scattered over the barren hillside. It may take many decades for forest regeneration to occur after this and in some localities it may never again be possible.

Stages in the degradation of forest are as follows:

Evergreen forest → maquis → garigue → steppe.

Regeneration may take place in the reverse order if man and his domestic animals cease their activities. The whole balance between natural, semi-natural and man-made plant communities is summarized in Fig. 5.

EVERGREEN FORESTS

Probably the most extensive forests in the Mediterranean region are those dominated by the **Holm Oak**. This tree grows on all kinds of soils and may occasionally form dense forests, but more commonly it occurs in scattered clumps. Holm Oak forests are rarely seen in their climax state, when they form very dense and dark woods up to 12–15 m. high with a shrub layer of Strawberry Tree, *Phillyrea*, *Rhamnus* and *Viburnum*, and many climbing plants such as *Clematis*, Honeysuckle, *Smilax* and Black Bryony. They may form impenetrable

6

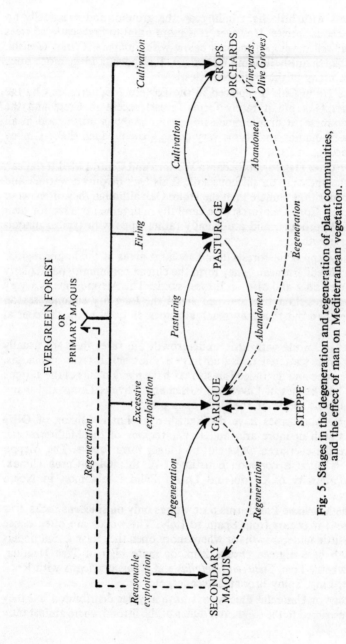

Fig. 5. Stages in the degeneration and regeneration of plant communities, and the effect of man on Mediterranean vegetation.

7

thickets with little light falling on the ground, and practically no herbaceous species. However, it is more usual to find scattered trees and a well developed maquis below with Junipers, *Cistus, Genista,* Lentisk, Spanish Broom, Rosemary, Lavender, and in more open places many of the garigue species.

The Holm Oak is replaced in the Eastern Mediterranean by the **Kermes Oak** but it rarely forms forests except in Crete and the Peloponnese; it occurs much more commonly as brushwood in all kinds of maquis and forms very dense growth which shelters many herbaceous species.

The **Cork Oak** forests occur in Western and Central Mediterranean regions and only on siliceous soils. This Oak favours a warmer and more maritime climate than the Holm Oak although they often occur together. Cork Oak forests are carefully maintained by man for their valuable product, and are usually rather open with typical maquis undergrowth.

The **Aleppo Pine** forests dominate large areas of the hotter regions of the Mediterranean. They form the climax community particularly on limestones and also on littoral sands. The Aleppo Pine is a very drought-resistant tree and can stand the long dry summer season better than most; it may reach altitudes of 1,000 m. in the coastal hills.

Dense forests with little undergrowth are rare; they are usually open with scattered trees and have a thick shrub layer of maquis. More open and degraded forests may have many plants of the garigue with low bushes of Lavender, Rosemary, Thyme, *Cistus* and many herbaceous species in the more open places.

Some authorities have distinguished a climatic-climax of **Olive** and **Carob** in more arid and hotter regions of the Mediterranean where the evergreen Oaks can no longer form forests. The Aleppo Pine is often a common constituent of this **Olive-Carob** climax. A community of Carob and Dwarf Palm is common in North Africa.

The **Maritime Pine** forms pure woods only on siliceous rocks near the sea; it occurs from Spain to Italy. The woods are often dense with little undergrowth, or where more open they have a rich undergrowth of evergreen shrubs 2 m. or more high of Tree Heather, Strawberry Tree, *Cistus salviaefolius* and *C. monspeliensis* with Rock Rose, Ling, Spiny Broom, Lavender, etc.

Stone or **Umbrella Pine** woods have a wider distribution but they are restricted to the sands and dunes of the littoral where almost pure

8

forests are formed with a luxuriant undergrowth, particularly of Phoenician Juniper, Lentisk, Spanish Broom, Rosemary, *Phillyrea*, *Cistus*, etc.

Cypress Woods are generally of scattered trees with an undergrowth of evergreen and deciduous shrubs and woolly-leaved members of the garigue. They occur in Crete, Rhodes and Cyprus.

Laurel Woods occur in Greece, Crete and the Balkans.

ALTITUDE ZONATION OF VEGETATION

The true Mediterranean vegetation of the **Evergreen forests** forms a broad or narrow zone around the Mediterranean littoral. In open sunny valleys it may penetrate many miles inland or cover practically the whole of a peninsula, such as the Peloponnese, or an island like Crete. Elsewhere, as for example in the Italian Riviera, it may be quickly replaced by a more temperate deciduous forest a mile or so from the sea. The altitude range of the evergreen forest will also depend on the local climate, but it may reach an altitude of 500 m. or more.

Above the Evergreen Forest Zone lies a **Deciduous forest** zone of deciduous Oaks, Ash, Hornbeam, Sweet Chestnut, Beech, from altitudes of 800–1,800 m. and above this a sub-alpine **Coniferous forest** zone of Silver Fir, Pines, and in certain places Cedar (see Fig. 6).

MAQUIS

The maquis, or macchie, is a very characteristic and natural type of Mediterranean plant community. It forms very dense and sometimes impenetrable thickets of tall shrubs, 2 m. or more high, with stiff densely twiggy branches and small dark green leathery leaves. In springtime it brightens the hillsides with splashes of colour, with the pink and white of the Cistuses, the yellows of the Brooms and the snow-like dusting of the myriads of flowers of the Tree Heather. During the remainder of the year it forms a dark green mantle over the hillsides, thickening in the valleys and thinning on the drier ribs of the hills. The dark green contrasts vividly with the bare rocks and the dried-up fields in summer, and in the burning heat of the day resinous aromatic oils are given off from the leaves of many maquis plants.

It is not possible to know whether the maquis is the highest expression of vegetative development, or climax, under certain conditions

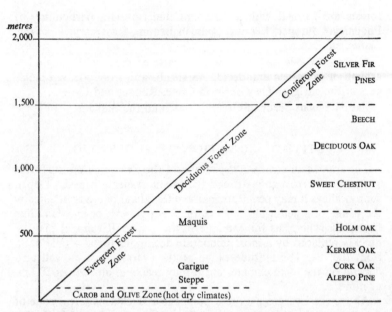

Fig. 6. Altitude zonation of the plant communities in the Mediterranean region.

in the Mediterranean. In some cases it probably is, in which case it is known as 'primary maquis', but in most cases it is undoubtedly the result of man's activity on the 'primaeval' evergreen forest, and is consequently called 'secondary maquis'.

There are so many useful products to be gathered from the maquis such as gums, resins, tannins, brushwood, charcoal, dyes and fibres, that man is continuously cutting and changing the natural vegetation; he often clears the vegetation and cultivates the ground, only to abandon it later, so that today the maquis shows a bewildering number of variations all grading into each other. The classification of maquis is difficult, and the following are some of the more widespread variants:

'High Maquis' is characterized by the presence of a number of trees; it may grow as high as 4–5 m. and include the Strawberry Tree and *Arbutus andrachne*, Holm and Kermes Oaks, Phoenician Juniper, Judas Tree, Olive and Aleppo Pine, as well as larger shrubs such as Myrtle, Tree Heather, Wig Tree, *Phillyrea media*, and Spanish Broom.

There are all gradations between this and 'Low Maquis' where the bushes are 1½–2 m. high and there are no tree-forming species. Commoner species are Lentisk, *Phillyrea media*, Rosemary, Jerusalem Sage, Butcher's Broom, *Erica* species, *Paliurus spina-christi*, *Cistus salviaefolius*, *C. monspeliensis* and *C. villosus*.

In open patches many herbaceous perennials grow including some of the most interesting bulbous, tuberous plants, and annuals.

Cistus maquis is a very widespread 'low maquis' in hot dry localities, with *Cistus monspeliensis* often dominating large areas, while *C. villosus* dominates extensively in the Eastern Mediterranean. It can stand heavy grazing and often develops on abandoned cultivated areas.

Mixed **Lentisk-Carob-Myrtle maquis** has many variations and occurs extensively on hot dry lower hills and coastal regions, particularly in the Eastern Mediterranean. It contains many common species like Spiny Broom, Terebinth, *Rhamnus alaternus*, Kermes Oak, Butcher's Broom and Hawthorn. With excessive cutting and grazing it may be reduced largely to rounded scattered bushes of Lentisk which harbour many herbaceous species.

There are many other specialized types of maquis.

GARIGUE

Garigue is an important type of Mediterranean vegetation as indicated by the different names given to it, namely tomillares (Spain), phrygana (Greece) and batha (Palestine). Extensive areas of the hottest and driest terrain are covered with garigue and it is easily distinguished by its low scattered bushes, rarely more than ½ m. high, dotted over the hillsides, with bare patches of rock, sand or stony ground between. Many of the shrubs are spiny and have small leathery heather-like leaves often covered with woolly grey hairs; many are aromatic. The species inhabiting this 'rock heath' are very numerous; there are woody and herbaceous perennials, biennials and annuals in abundance, e.g. the Greek phrygana has over 200 different species. Many of our culinary 'herbs' such as Thyme, Rosemary, Sage, Savory, Hyssop, Lavender, Garlic and Rue come from the garigue, while it is also the home of many ornamental bulbous and tuberous plants like Tulips, Crocuses, Irises, Grape Hyacinths, Fritillaries, Star of Bethlehem, Garlics, *Orchis* and *Ophrys* species. It is largely plants of the garigue which give so much colour to the Mediterranean spring landscape, but the sudden burst of flowering in April is soon followed by a dry and parched countryside with dead fruit heads and dry dusty-leaved shrubs.

Typical plants of the Eastern garigue are the Spiny Burnet, *Corido-thymus capitatus*, *Thymelaea tartonraira* and *T. hirsuta*, *Genista acanthoclados*, *Euphorbia acanthothamnos*, *Hypericum empetrifolium* and some shrubs of the maquis such as the Kermes Oak and *Cistus* species.

Each maquis community degenerates into a different garigue, and there are a great many forms of garigue. The 'tomillares' of Spain refer to the 'Thyme' and other members of the *Labiatae* which are so common, such as Lavender, Sage, *Phlomis*, etc. In contrast to forest and maquis, local variations of the soil have a much more profound effect on the composition of the garigues; many are developed on calcareous soils.

It is of interest to note that both the maquis and garigue are related to the 'heaths' of Western Europe and Great Britain. Quite extensive tracts of 'heath garigue' with the Ling, *Calluna vulgaris*, as the dominant plant, are found in Italy, and the Gorse, *Ulex europaeus*, is another species of some importance in the Western maquis.

STEPPES

Where man has cut down the natural vegetation, or his animals have grazed it away, and where the soil has been destroyed and the underlying rock exposed, after a period of time 'steppes' may develop. Shrubs, which play such an important part in building up the soil and protecting other herbaceous species, are no longer present. In their place many annuals, and herbaceous perennials with deep root systems or with swollen storage organs below ground, establish themselves. In the dry hot summer the aerial stems die and underground parts remain dormant until the return of the rains in autumn. The flowering season is very short and there is often a brilliant show of colour. Many members of the important 'Mediterranean' families are present, including the Pea, Mint, Mustard, Pink, Buttercup, Lily and Grass families, as well as Thistle-like members of the Daisy family and members of the Carrot family such as *Ferula communis*, and genera such as the Mulleins.

Some steppes are predominantly grassy with many kinds of annual grasses, as well as Medicks, Clovers and Plantain species. Others have a mixture of perennial plants including Anemones, Irises, Poppies, and members of the Daisy and Borage families, as well as many bulbous plants.

Asphodel Steppe may occur in cases of extreme soil depauperation when the tall and striking *A. microcarpus* and *A. albus*, and the enormous bulbs of the Sea Squill, become abundant.

In addition there are distinctive assemblages of plants in special localities in the Mediterranean region. They often contain plants with a much wider distribution as well as plants that are exclusively Mediterranean. Plants found along the Mediterranean littoral on cliffs, sands, dunes and salt marshes are often widespread in Britain in similar localities, as for example the Sea Holly, Glassworts and Sea Rocket.

Fresh-water and marsh habitats are not common in the Mediterranean but they also contain species such as Water Crowfoots, Pondweeds and Reeds which are common in Britain.

Many weeds of cultivation are now becoming cosmopolitan and quite a number of weed plants which are casual in Britain originate in the Mediterranean region.

CULTIVATED AND ORNAMENTAL PLANTS

Since man has been on the move in the Mediterranean, there has been an ever-increasing influx of plants from other regions. The Greeks and the Phoenicians were probably the first to transport plants: Olives, Figs and Pomegranates in all probability had their origin in the Orient, yet by the end of the pre-Christian era they were cultivated throughout the whole Mediterranean region. The Arabs brought the Orange from China.

During the last 300 years this rate of arrival of new plants from all parts of the world has been enormously accelerated. Plants have been brought in for ornament, such as the quick-growing Eucalyptuses and Acacias ('Mimosas') from Australia and the familiar 'boulevard' Palm from the Canary Islands. The American Agave, or Century Plant, of sub-tropical America, and the Prickly Pear said to have been brought to the Mediterranean by Christopher Columbus from an unknown tropical country, have become very familiar features of the Mediterranean landscape. Pepper Trees, Bougainvillea, False Acacia and Persian Lilac are now commonly planted in every small town or village.

Fruits and vegetables have also been brought from overseas to add to the already rich variety of Mediterranean harvests, such as the Loquat from Japan, Lemons and Tangerines, Capsicums and Chillies, and not least Maize, Potatoes and Tomatoes from the 'New World'.

At the same time many weeds have spread from their ancestral countries and are becoming widespread in the Mediterranean; plants like the Bermuda Buttercup from South Africa and Spiny Clotbur from America are two distinctive examples.

The vegetation of the Mediterranean is undergoing rapid change; even the slow-moving evolutionary changes are being affected by the mixing up of previously isolated races of plants; interbreeding and hybridization are encouraged. The formation of new habitats by man, and the great influx of foreign plants has already brought about such considerable changes in the last 2,300 years that, should Theophrastus visit the shores of the Mediterranean again, he would find it necessary to add several more 'books' to his 'Enquiry into Plants'.

GLOSSARY OF TERMS

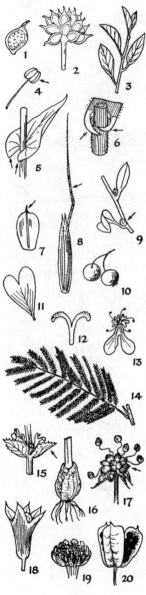

achene A one-seeded ovary (1); there are usually many achenes on a fruiting head (2), as in buttercup.

adpressed, appressed Pressed flat to a surface; commonly used of flattened hairs (see hair).

alien Believed to have been introduced by man and now more or less naturalized.

alternate Leaves placed singly at different heights on a stem (3).

annual A plant completing its life and seeding in one year or less.

anther The part of the stamen containing the pollen grains (4).

apex, apical Top-most point, or pointed end; hence apical bud or flower.

ascending Rising upwards at an angle.

auricle An ear-like flap of tissue at the base of a leaf (5), curiously shaped in grasses (6).

awl-shaped Broad-based and tapering to a sharp point (7).

awn A long stiff bristle-like projection borne at the end or from the side of an organ (8).

axil, axillary The angle between the leaf and stem; hence axillary flower or bud (9).

berry A fleshy rounded fruit usually with hard pips or seeds (10).

biennial A plant which grows and develops in the first year, and fruits and seeds in the second.

bifid Divided deeply into two as far as the middle (11).

bifurcate Divided into two equal branches arising from the same point (12).

bipartite Divided very deeply into two, to below the middle (13).

bipinnate Twice cut; of leaves cut into segments which are themselves cut (14).

blade The flattened part of an organ, such as a leaf or petal.

bract A little leaf or scale-like structure from the axil of which a flower often arises (15).

bulb A swollen underground bud-like structure remaining dormant below ground during unfavourable growth periods (16).

bulbil A small bulb or tuber arising from the axil of the leaf or among the flowers, and reproducing the plant (17).

calyx The sepals collectively; often joined together in a tube, the calyx tube (18).

capitulum A head of small stalkless flowers crowded together at the end of the stem, as in the daisy family (19).

capsule A dry fruit, of two or more carpels, which splits open when ripe (20).

15

carpel One of the units of the female part of the flower; they are either separate or fused together into a fruit (21, and in transverse section, 22).

cartilaginous Tough and hard.

casual A foreign plant introduced into a country, which appears periodically but cannot maintain itself for long.

catkin A crowded spike of tiny flowers, usually hanging and tassel-like (23).

cell Of a fruit, the cavity containing the seeds.

ciliate Fringed with hairs along the margin (24).

column In orchids, the combination of stamens and stigma.

composite A member of the daisy family (25).

compound Two or more similar parts in one organ; hence a compound leaf has two or more separate leaflets (q.v.).

cone A distinct rounded or elongate structure composed of many overlapping scales which bear pollen or seeds when ripe (26).

corm A swollen underground stem surrounded by scales, the tunic, and replaced annually by a new corm (27).

corolla The petals collectively; often joined together into a tube, the corolla tube (28).

corona or **crown** Structures or appendages which stand out from the petals and together form a ring round the centre of the flower (29).

crucifer A member of the mustard family.

cyme A broad, more or less flat-topped flower cluster, with the central flowers opening first (30).

decurrent When the blade or stalk of the leaf continues, as a rib, down the side of the stem (31).

dehiscent Splitting; particularly of fruits to allow the seeds to escape (32).

dichotomous Divided into two equal forks, and often forked again and again (33).

disk A fleshy part of the receptacle which surrounds or surmounts the ovary (34).

disk-florets The tube-like flowers at the centre of the flower heads of some members of the daisy family (35).

downy Covered with short, weak, soft hairs.

drupe A fleshy fruit with an inner hard stone enclosing the seeds (36).

egg-shaped With an outline broader towards the base than the apex, and round-ended (37).

elliptic Oval and narrowed to rounded ends (38).

endemic Native of only one country or area.

entire Whole, without teeth, lobes or indentations.

epicalyx A calyx-like structure outside, but close to the true calyx (39).

eye The centre of the flower when distinct from the remainder (40).

falls The outer set of petals which are usually broader than the inner, and often drooping, as in iris (41).

family A group of plants with many common characteristics; usually composed of many genera.

feathery Cut into many fine segments (42).

female flowers Flowers with an ovary but without stamens.

fertile Capable of producing fruit.

filament The thread-like stalk of the stamen bearing the anthers (43).

floret A small flower, usually one of a dense cluster.

form, forma A slight but distinctive variant of a species, often occurring sporadically.

free Not joined to other organs, e.g. as petals to stamens, or petals to each other.

garigue A type of vegetation composed of low, scattered, often spiny and heath-like shrubs, with usually patches of bare ground between the clumps of shrubs.

genus (plural genera) A classificatory term for a group of closely related species: a number of genera form a family. The generic name is the first part of the Latin binomial, e.g. *Ranunculus* (genus) *arvensis* (species).

glabrous Not hairy, and hence generally smooth.

gland Organs of secretion usually on the tips of hairs; hence glandular hairs (see hair).

glaucous Covered or whitened with a bloom which is often waxy, thus giving the organ a bluish or greyish colour.

glume A chaff-like bract; in particular the bracts at the base of the spikelets of grasses (44).

granular Covered with very small grains; minutely or finely mealy.

hair A fine projection from a surface; one-celled or many-celled, simple or branched, straight or curved, etc. Special kinds of hair include glandular (45), crisped (46) and star-shaped (47) hairs. Long, soft hairs are described as woolly (48) and those pressed flat to a surface as adpressed (49).

head Of flowers or fruits, crowded together at the end of a common stalk (50).

herb A plant which has no woody stem and is soft and leafy; also a plant used in seasoning or medicine.

herbaceous Non-woody, soft and leafy. Of a plant organ, having the soft texture and green colour of leaves.

hermaphrodite With the stamens and ovaries present in the same flower.

hoary Covered with close white hairs.

hybrid A plant resulting from cross-breeding between two different species, and with some characteristics of each parent.

inferior Lower or below; hence, inferior ovary, where the ovary is situated below the calyx (51).

interrupted Not continuous.

introduced Not native; a plant brought into a country or region from elsewhere.

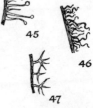

17

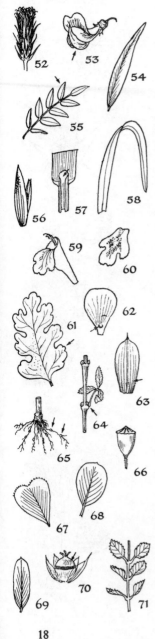

invested Covered or enclosed.

involucre A collection of bracts or leafy structures surrounding a flower head, groups of flowers, or a single flower (52).

keel A sharp edge on an organ resembling the keel of a boat. Also the lower petals of the flowers of the pea family (53).

lanceolate Shaped like a lance head; several times longer than wide, broadest towards the base and narrowed to the tip (54).

lateral Borne on the side of an organ.

leaflet The individual part of a compound leaf which is usually leaf-like with a stalk of its own (55).

leaf-stalk The stalk of a leaf which bears the blade.

lemma In grasses, the glume or bract which bears the flower; thus the lower of the two bracts of the flower (56)

ligule A small projection at the junction of the blade of a leaf with its sheath, found commonly in the grasses (57).

linear Long and narrow with nearly parallel sides (58).

lip One of two flap-like projections of a calyx or corolla (59); in orchids, the lowest of the three petal-like segments (60).

lobe A part or segment of an organ deeply divided from the rest of the organ but not separated (61).

male flower A flower containing stamens but no ovary.

maquis A type of vegetation composed of tall shrubs and often scattered trees. It may form a dense thicket.

monocarpic Flowering only once and then dying.

mucilage A viscid, slimy, gum-like substance; often produced in ducts or canals in the plant.

native A plant naturally occurring in an area and not introduced from elsewhere.

naturalized Thoroughly established in an area, but originally coming from another region.

nectary A gland which gives out a sugary liquid and serves to attract insects (62).

nerves Prominent veins, e.g. in a leaf or petal (63).

node A point on the stem where one or more leaves arise (64).

nodule A small rounded swelling, as on the roots of members of the pea family (65).

nut A one-seeded fruit with a hard outer covering.

ob- Inverted; used as a prefix: thus, *obconical* (66), inverted conical; *obcordate* (67), inverted heart-shaped; *obovate* (68), egg-shaped with the broadest end towards the tip, in contrast to ovate, where the broadest part is towards the base.

oblong An elongated but relatively wide shape, e.g. of leaf, with parallel sides (69).

operculum A lid or cover, e.g. of a fruit (70).

opposite Of two organs, arising at the same level on opposite sides of the stem (71).

orbicular Rounded in outline with length and breadth about the same. (72)

ovary The part of the flower containing the ovules and later the seeds (73).

ovate With an outline like that of a hen's egg with the broadest part towards the base (74).

ovule A structure, containing the egg, which after fertilization becomes the seed.

palate A rounded projection on the lower lip of a two-lipped corolla which more or less closes the opening or throat (q.v.) of the corolla, as in snapdragon (75).

palea, pale In grasses, the upper of the two bracts which enclose the flower (76).

palmate Lobed or divided in a palm-like or hand-like manner (77).

panicle A branched cluster of stalked flowers in which the oldest flowers are towards the base (78).

pappus Hairs or bristles which replace the calyx in some members of the daisy family, and often facilitate fruit dispersal (79).

parthenogenetic Of seeds, developing without fertilization.

perennating Surviving the winter, or unfavourable growth period.

perennial Living for more than two years and usually flowering each year.

perianth The outer non-sexual parts of the flower; usually composed of petals (p) and sepals (s) (80).

perianth segment An individual petal or sepal; commonly used when petals and sepals are indistinguishable as either, as in the lily family.

petal One of the inner set of organs surrounding the sexual parts of the flower; usually brightly coloured and conspicuous (see also perianth).

phyllode Leaf-stalk which has become flattened and leaf-like, usually replacing the true leaves (81).

pinnate The arrangement of leaflets in two rows on either side of the stalk (82).

pinnately lobed A leaf with opposite pairs of lobes; not with separate leaflets (83).

pod A general term for a dry splitting fruit, formed from one unit or carpel.

pollen Small grains which contain the male reproductive cells.

pome A fruit, such as an apple, in which the fleshy part is formed from the receptacle, the core is the wall of the ovary, and the pips are the seeds.

radical Commonly used of leaves which arise from the rootstock.

ray One of the stalks of an umbel (84).

ray floret The strap-shaped floret of many members of the daisy family (85).

19

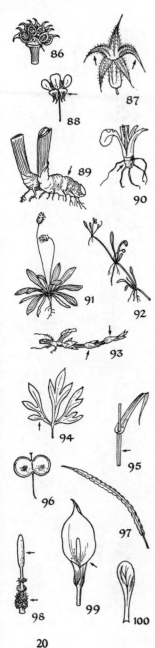

receptacle The uppermost part of the flower stalk which bears the parts of the flower, or the florets in the daisy family (86).

recurved Bent backwards or downwards in a curve (87).

reflexed Bent abruptly backwards or downwards (88).

rhizome A creeping underground stem which sends up new leaves and stems each season (89).

rhomboidal Roughly diamond-shaped.

rootstock A short erect underground stem (90).

rosette An arrangement of leaves radiating from a crown or centre, usually at the surface of the ground (91).

runner A trailing stem which roots at the nodes, forming new plants which eventually become detached from the parent (92).

saprophyte A plant which derives its food wholly or partially from dead organic matter.

scale Any thin, dry flap of tissue; usually a modified or degenerate leaf (93).

scrub Any plant community dominated by shrubs; as thicket.

seed pod An ovary containing seeds.

segment One of the parts of a cut or partially divided leaf, corolla or calyx (94).

sepal One of the outer set of usually green organs surrounding the flower bud. The sepals collectively form the calyx; they may sometimes be brightly coloured and petal-like (see also perianth).

sheath A more or less tubular structure surrounding another; as in the lower part of the leaves of grasses (95).

siliceous Soil or rock containing silica, and in consequence becoming acid. Sandstones are rich in silica (or quartz) and are siliceous.

silicula A pod-like fruit of the mustard family which is not more than twice as long as broad, and more often broader than long (96).

siliqua A pod-like fruit of the mustard family which is longer, often many times so, than broad (97).

silky Having a covering of soft fine hairs.

simple Of a leaf, not divided up into segments; of stems, unbranched.

sinus The cleft or recess between two lobes.

spadix A fleshy spike bearing flowers and often ending in a swollen club-like apex; as in arum (98).

spathe A large bract enclosing a flower head; conspicuous in arum where it is coloured (99).

spathulate Spoon or paddle-shaped; broader towards the apex and narrowed to the base (100).

species (abbreviated **sp.**) A group of individuals having similar characteristics, and which interbreed. Species are grouped together into genera. The specific name is the second part of the Latin binomial.

spike A close elongated head of more or less stalkless flowers (101).

spikelet A secondary spike or part of a compound spike. Also in grasses a group of one or more flowers subtended by two sterile bracts or glumes (102).

spreading Standing outwards or horizontally.

spur A hollow, more or less cylindrical projection from a petal or sepal; it usually contains nectar (103).

stamen One of the male reproductive organs of the flower which bears the pollen (104).

staminode An infertile or rudimentary stamen without pollen (105).

standard The broad upper petal of a flower of the pea family (106). Also the inner petals of an iris flower which stand erect.

stem The main axis of a plant, which is both leaf-bearing and flower-bearing.

steppe A treeless landscape. A type of vegetation consisting of grasses, other herbaceous plants but no shrubs which, in the Mediterranean region, develops with excessive grazing, absence of soil cover and low rainfall.

sterile Lacking functional sex organs.

stigma The part of the female organ which receives the male pollen (107). Normally situated at the top of the style, it is often covered with a sticky secretion.

stipule A scale-like or leaf-like appendage at the base of the leaf-stalk; usually paired (108).

style A more or less elongated projection of the ovary which bears the stigma (109).

subspecies (abbreviated ssp.) A group of individuals within a species which have some distinctive characteristics, and often a well marked geographical range. The characteristics separating subspecies are more marked than those separating varieties.

subspontaneous Of a plant, seeding and establishing itself in the wild after being brought into a locality from elsewhere.

subtending To stand below and close to, e.g. a bract bearing a flower in its axil.

sucker A shoot arising below ground, often at some distance from the main stem (110).

taproot The main descending root.

tendril A slender, clasping, twining organ, often formed from a leaf or part of a leaf (111).

tessellated Chequered or netted.

throat The opening or orifice of a tubular or funnel-shaped corolla, or calyx (112).

throat-boss A swelling in the throat: see palate.

trefoil With three lobes or leaflets; as a clover leaf (113).

trifoliate Having three leaflets: trefoil.

tube The fused part of the corolla or calyx (114).

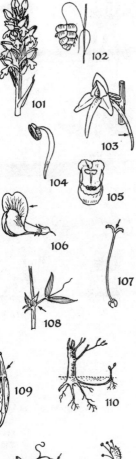

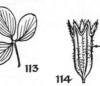

21

tuber A swollen part of a stem or root, formed annually and usually underground (115); also used for a perennial underground swollen stem, as in cyclamen.

umbel A cluster of flowers whose spreading stalks arise from the apex of the stem, resembling the spokes of an umbrella (116).

umbellifer A member of the carrot family with flowers in umbels.

undershrub A woody perennial, rarely as tall as one metre high, generally less, and often more or less prostrate.

undulate Wavy in a plane at right angles to the surface (117).

variety (abbreviated var.) A group of plants within a species with one or more distinctive characteristics, e.g. a marked colour variation, unusual leaf form, etc., etc. Varieties may have a distinct geographical range.

valve One of the parts into which a capsule splits.

veins Strands of strengthening and conducting tissues running through the leaf, and other organs (118).

viviparous Sprouting and germinating on the parent plant (119).

wavy With regular curved indentations in the same plane as the surface (120).

wedge-shaped Narrowest at the point of attachment and increasing regularly in width to the apex (121).

whorl More than two organs of the same kind arising from the same level; thus whorled (122).

wing A dry thin expansion to an organ (123). Also the lateral petals of the flowers of the pea family.

woolly With long, soft, more or less tangled hairs.

ABBREVIATIONS

Circum-Medit. = Whole Mediterranean basin but not necessarily in Egypt and Libya where the climate is more Saharan.

cm. = centimetre (0·39 in.)
m. = metre (39·37 in.)
mm. = millimetre (0·04 in.)
sp. = species
ssp. = sub-species
var. = variety

Abbreviations after the Latin names of plants refer to the author of that name, e.g. Boiss. = Boissier; L. = Linnaeus; Hal. = Halácsy; DC. = de Candolle.

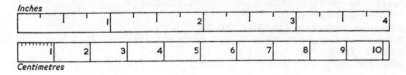

THE COLOUR PLATES

1 Juniperus communis × 1/2

2 Juniperus phoenicea × 1/3

3 Pinus halepensis × 1/3

4 Pinus pinaster × 1/6

5 Abies cephalonica × 1/8

6 Quercus suber × 1/12

7 Ficus carica × 1/6

8 Aristolochia sempervirens × 4/5

9 Aristolochia rotunda × 4/5

10 Cytinus hypocistis × 2/3

11 Bougainvillea spectabilis × 1/4

12 Carpobrotus acinaciformis × 1/4

13 Carpobrotus edulis × 1/2

14 Lampranthus roseus × 1/5

15 Silene succulenta × 1/4

16 Silene colorata × 1/7

17 Saponaria calabrica × 1/2

18 Kohlrauschia glumacea × 1/2

19 Helleborus cyclophyllus × 1/3

20 Delphinium halteratum × 1/3

21 Anemone blanda × 1/3

22 Anemone hortensis stellata × 1/3

23 Anemone hortensis × 1/6

24 Anemone coronaria × 1/2

25 A. hortensis fulgens × 1/2 26 A. coronaria × 1/2 27 A. coronaria × 1/3

28 Paeonia broteroi × 1/2

29 Ranunculus ficaria grandiflora × 1/3

30 Ranunculus asiaticus × 1/2

31 Laurus nobilis × 1/2

32 Adonis annua × 2/3

33 Bongardia chrysogonum × 1/4

34 Papaver rhoeas × 1/3

35 Glaucium corniculatum × 1/2

36 Hypecoum procumbens × 1

37 Fumaria capreolata × 1/4

38 Malcolmia maritima × 1/4

39   Moricandia arvensis × 1/4

40   Alyssum saxatile × 1/5

42   Matthiola sinuata × 2/5

41   Lobularia maritima × 1 1/4

43   Aethionema saxatile graecum × 1/2

44 Reseda lutea × 1/5

45 Reseda alba × 1

46 Platanus orientalis × 1/4

47 Eriobotrya japonica × 1/12

48 Poterium spinosum × 1/3

49 Anagyris foetida × 2/5

50 Acacia cyanophylla × 1/2

51 Cytisus triflorus × 1/3

52 Lupinus angustifolius × 1/2

53 Lupinus hirsutus × 1/3

54 Cercis siliquastrum × 1/4

55 Calycotome infesta × 1/4

56 Spartium junceum × 1/4

57   Genista cinerea × 1/40

58   Genista equisetiformis × 1/2

59   Genista hirsuta × 1/3

60   Genista acanthoclados × 3/4

61   Ononis speciosa × 1/5

62   Medicago marina × 1/2

63 Trifolium uniflorum × 1/2

64 Medicago arborea × 4/5

65 Trifolium stellatum × 1

66 Trifolium purpureum × 1/3

67 Anthyllis vulneraria spruneri × 1/6

68 Coronilla emerus emeroides × 1/2

69 Anthyllis cytisoides × 1/8

71 Coronilla juncea × 2/3

70 Physanthyllis tetraphylla × 1/2

72 Tetragonolobus purpureus × 3/4

73 Psoralea bituminosa × 1/3

74 Hedysarum coronarium × 1/2

75 Vicia hybrida × 1/2

76 Pisum elatius × 1/2

77 Lathyrus aphaca × 3/5

78 Lathyrus tingitanus × 1

79 Oxalis pes-caprae × 1/3

80 Ruta chalepensis × 1

81 Oxalis articulata × 1/2

82 Geranium molle grandiflorum × 2/3

83 Geranium tuberosum × 1

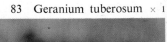

84   Linum campanulatum × 1/3

85   Linum pubescens × 3/5

86   Melia azedarach × 1/4

87   Melia azedarach × 1/3

88   Euphorbia spinosa × 1/3

89   Euphorbia acanthothamnos × 1/8

90 Euphorbia dendroides × 1/15

91 Euphorbia veneta × 1/20

92 Euphorbia biglandulosa × 1/9

93 Euphorbia characias × 1/3

94 Euphorbia myrsinites × 1/3

95 Pistacia terebinthus × 1/2

96 Pistacia terebinthus × 2/5

97 Pistacia lentiscus × 4/5

98 Malva sylvestris × 1/2

100   Cistus villosus × 2/5

99   Cistus albidus × 4/5

102   Cistus salviaefolius × 1/2

101   Cistus villosus creticus × 4/5

103   Cistus monspeliensis × 1/3

104 Cistus ladaniferus × 1/5

105 Cistus ladaniferus × 1/2

106 Cistus crispus × 2/5

107 Cistus populifolius × 2/5

108 Halimium atriplicifolium × 1/2

109 Tuberaria guttata × 1/2

110 Opuntia ficus-indica × 1/20

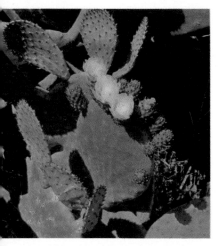

111 Opuntia ficus-indica × 1/8

112 Thymelaea hirsuta × 2/3

113 Daphne gnidium × 2/3

114 Thymelaea tartonraira × 1

115 Ferula communis × 1/24

116 Ferula chiliantha × 1/15

117 Tordylium apulum × 1

118 Smyrnium perfoliatum × 1/5

119 Arbutus unedo × 1/2

120 Arbutus andrachne × 1/3

121 Erica arborea × 1/15

122 Coris monspeliensis × 1 1/4

123 Cyclamen repandum (Rhodes) × 1/3

124 Cyclamen neapolitanum × 1/3

125 Cyclamen orbiculatum × 4/5

126 Cyclamen persicum × 1/3

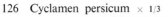

127　Anagallis monelli × 1/2

128　Anagallis arvensis × 2/3

129　Nerium oleander × 4/5

130　Vinca major × 3/5

131　Vinca difformis × 1/2

132 Limonium sinuatum × 1/5

133 Convolvulus tricolor × 2/3

134 Styrax officinalis × 3/4

135 Convolvulus elegantissimus × 1/4

136 Convolvulus althaeoides × 1/2

137 Convolvulus cantabricus × 2/5

138 Convolvulus lanuginosus × 2/3

139 Borago officinalis × 2/5

140 Lycopsis variegata × 4/5

141 Echium lycopsis × 1/3

142 Echium diffusum × 1/2

143 Anchusa hybrida × 4/5

144 Anchusa azurea × 1/3

145 Echium judaicum × 1/3

146 Onosma frutescens × 1/3

147 Lithospermum purpureo-coeruleum
× 1

148 Cerinthe major × 2/5

149 Cerinthe major purpurascens × 2/3

150 Lithospermum diffusum × 1 1/4

151 Cerinthe retorta × 1

152 Alkanna orientalis × 1

153 Lavandula stoechas × 3/4

154 Lamium moschatum × 1/2

155 Salvia triloba × 2/5

156 Salvia verbenaca × 2/3

157 Satureia thymbra × 4/5

158 Micromeria nervosa × 1 1/5

159   Ajuga chamaepitys × 2

160   Teucrium fruticans × 1

161   Rosmarinus officinalis × 1/4

162   Phlomis purpurea × 1/4

163   Phlomis fruticosa × 1/15

164   Phlomis lychnitis × 1/4

165 Hyoscyamus albus × 1/2

166 Hyoscyamus aureus × 1/2

167 Solanum sodomaeum × 1/4

168 Datura metel × 1/8

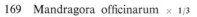

169 Mandragora officinarum × 1/3

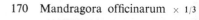

170 Mandragora officinarum × 1/3

171 Antirrhinum latifolium × 1/4

172 Antirrhinum orontium × 5/8

173 Linaria triphylla × 3/8

174 Linaria hirta × 1 1/2

175   Bellardia trixago × 1

176   Verbascum undulatum × 1/1

178   Scrophularia peregrina × 2/3

177   Verbascum undulatum × 1/7

179   Globularia alypum × 1/3

80 Orobanche ramosa × 2/3

181 Orobanche crenata × 1/10

82 Plantago indica × 1/3

183 Putoria calabrica × 1

184　Centranthus ruber × 1/2

185　Fedia cornucopiae × 1

187　Scabiosa prolifera × 4/5

186　Viburnum tinus ×

188　Ecballium elaterium × 1

189 Campanula rupestris anchusaeflora × 1/3

191 Campanula ramosissima × 1 1/2

192 Legousia speculum-veneris × 1

190 Campanula rupestris × 1/4

193 Campanula drabifolia × 1/2

194 Helichrysum sanguineum × 3/5

195 Helichrysum stoechas × 1

196 Phagnalon rupestre × 1/3

197 Odontospermum maritimum × 1/3

199 Chrysanthemum coronarium
× 1

198 Chrysanthemum segetum × 2/5

200 Chrysanthemum coronarium
discolor × 1/2

201 Cladanthus arabicus × 3/4

202 Pallenis spinosa × 1/2

203 Anthemis chia × 1/2

204 Calendula arvensis × 1

205 Echinops ritro × 1/3

206 Notobasis syriaca × 1/2

207 Silybum marianum × 1/2

208 Galactites tomentosa × 1/6

209　Leuzea conifera × 2/3

210　Carthamus arborescens × 2/5

211　Carthamus lanatus × 2/5

212　Centaurea pullata × 1

213　Scolymus hispanicus × 2/5

214 Tolpis barbata × 3/4

215 Tragopogon porrifolius × 1 1/4

216 Crepis rubra × 1/2

217 Andryala integrifolia × 1/3

218 Launaea anthoclada × 1/4

219 Chamaerops humilis × 1/8

220 Arum italicum × 1/2

221 Arisarum vulgare × 1/2

223 Dracunculus vulgaris × 1/5

222 Biarum tenuifolium × 1/3

224 Arum dioscoridis × 2/5

225 Tulipa boeotica × 1/4

226 Tulipa sylvestris × 4/5

227   Fritillaria messanensis   × 1/2

228   Fritillaria graeca   × 3/4

230   Colchicum autumnale   × 1/3

229   Fritillaria acmopetala   × 3/4

231   Fritillaria libanotica   × 1/8

232 Asphodelus albus × 1/3

233 Asphodelus microcarpus × 1/20

234 Ruscus aculeatus × 3/4

235 Asphodeline lutea × 1/6

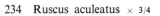

236 Gagea arvensis × 2/3

237 Ornithogalum montanum × 1/3

238 Asphodelus fistulosus × 1/4

239 Lloydia graeca × 1/2

240 Scilla bifolia × 3/5

241 Ornithogalum nutans × 1/2

242 Bellevalia ciliata × 2/5

243 Scilla peruviana × 1/5

244 Scilla hyacinthoides × 1/3

245 Muscari comosum × 1

246 Muscari racemosum × 1

247 Muscari commutatum × 2/3

248 Allium triquetrum × 1

249 Allium subhirsutum × 1

250 Allium neapolitanum × 1/2

251 Allium roseum × 1/6

252 Aphyllanthes monspeliensis × 2/3

253 Sternbergia lutea × 1/2

254 Sternbergia sicula × 1

255   Agave americana × 1/80

256   Gladiolus communis × 1/4

257   Narcissus tazetta × 1/2

258   Narcissus papyraceus × 1/3

259 Iris florentina × 1/5

260 Iris pumila attica × 1/2

261 Iris pumila attica × 1/2

263 Iris xiphium × 2/3

262 Hermodactylus tuberosus × 1

265 Iris chamaeiris × 2/3

264 Iris cretica × 1

266 Iris chamaeiris × 1/3

267 Iris histrio × 2/3

268 Romulea bulbocodium × 1

269 Iris sisyrinchium × 1

270 Crocus pallasii cartwrightianus × 2/3

271 Crocus flavus × 2/3

272 Ophrys argolica × 2 1/4

273 Ophrys lutea lutea × 1 3/4

274 O. fusca fusca × 2/3

275 O. fusca iricolor × 2

276 O. fusca omegaifera × 2/5

277 Ophrys speculum × 2 1/2

278 Ophrys scolopax cornuta × 1 1/3

279 O. sphegodes spruneri
　　　　　　　　× 1 1/2

280 O. sphegodes litigiosa
　　　　　　　　× 2

281 O. sphegodes
　　　mammosa × 3/5

282 O. bombyliflora × 2

283 O. fuciflora maxima
　　　　　　　　× 1 1/3

284 O. scolopax scolopax
　　　　　　　　× 1 2/5

285 O. scolopax attica × 2

286 Ophrys cretica × 1/4

287 Ophrys ferrum-equinum × 1 1/2

288 Limodorum abortivum × 1/3

291 Himantoglossum longibracteatum
× 3/8

289 Ophrys tenthredinifera × 1 3/4

290 Ophrys apifera × 1 3/4

292　Orchis papilionacea × 1/3

293　Orchis purpurea × 1

294　Orchis italica × 1

295　Orchis simia × 1

296 Orchis tridentata ×2/3   297   Orchis lactea × 3/4   298   Orchis collina × 1/2

299 Orchis laxiflora ×1/3   300   Orchis mascula olbiensis × 1/2   301   Orchis anatolica × 1/2

302 Orchis quadripunctata
× 1/2

303 Dactylorchis romana
× 1/2

304 Dactylorchis romana
× 1/2

305 Orchis provincialis
pauciflora × 3/4

306 Aceras anthropophorum
× 1/2

307 Anacamptis
pyramidalis × 3/4

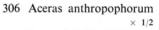

308 Serapias cordigera × 1

309 Serapias neglecta × 3/4

310 Serapias pseudocordigera × 3/4

311 Serapias lingua × 1 1/4

THE LINE DRAWINGS

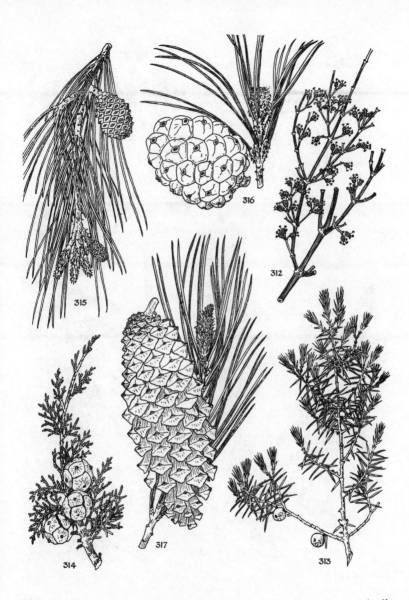

Plate 1   ($\times \frac{1}{2}$)
312 Ephedra fragilis ssp. campylopoda.   313 Juniperus oxycedrus.
314 Cupressus sempervirens.   315 Pinus halepensis.   316 P. pinea.
                       317 P. pinaster.

27

Plate 2   ($\times \frac{1}{2}$)
318 Quercus ilex. 319 Q. suber. 320 Q. coccifera. 321 Ficus syco-
        morus. 322 Amaranthus retroflexus.

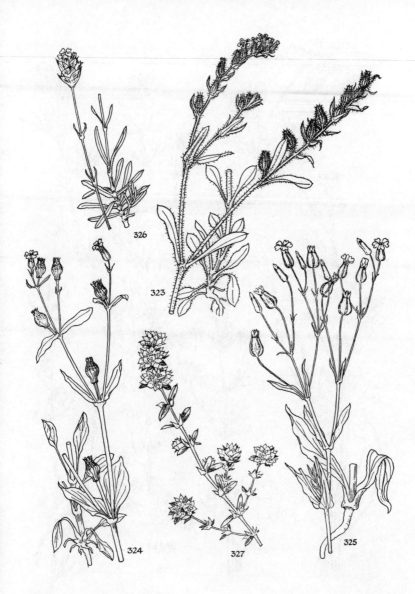

Plate 3   ($\times \frac{1}{2}$)
323 Silene gallica var. quinquevulnera. 324 S. behen. 325 Vaccaria pyrami-
     data. 326 Kohlrauschia velutina. 327 Paronychia argentea.

Plate 4 (× ⅓)

328 Nigella arvensis.   329 Delphinium peregrinum.   330 Clematis cirrhosa. 331 Ranunculus orientalis. 332 R. muricatus. 333 R. arvensis.

Plate 5

(× ½)

334 Bunias erucago.   335 Malcolmia maritima.   336 Matthiola longi-
      petala ssp. bicornis.   337 Biscutella laevigata.   338 Eruca sativa.

Plate 6 (× ½)

339 Acacia longifolia.   340 A. dealbata.   341 Ceratonia siliqua.
342 Colutea arborescens.

32

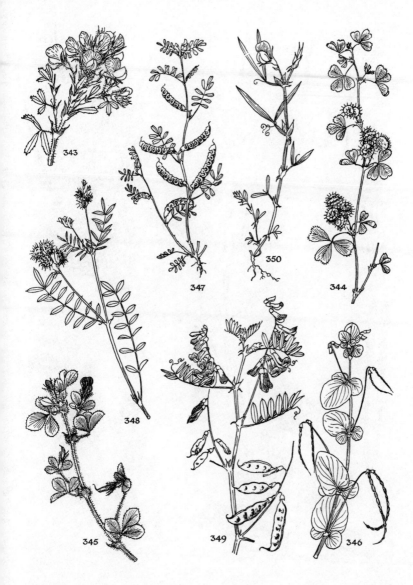

Plate 7 (× ½)
343 Ononis natrix.   344 Medicago polymorpha.   345 Tetragonolobus
purpureus.   346 Coronilla scorpioides.   347 Hippocrepis unisiliquosa.
348 Onobrychis caput-galli.   349 Vicia dasycarpa.   350 Lathyrus cicera.

Plate 8 (× ½)
351 Osyris alba. 352 Geranium rotundifolium. 353 Erodium gruinum.
354 Linum strictum. 355 Tribulus terrestris.

Plate 9 $(\times\frac{1}{2})$
356 Coriaria myrtifolia.   357 Cotinus coggygria.   358 Rhus coriaria.
359 Acer monspessulanum.   360 Paliurus spina-christi.   361 Zizyphus
spina-christi.

Plate 10 (× ½)
362 Schinus molle.   363 Rhamnus alaternus.   364 Punica granatum.
365 Sedum stellatum.   366 Erica verticillata.

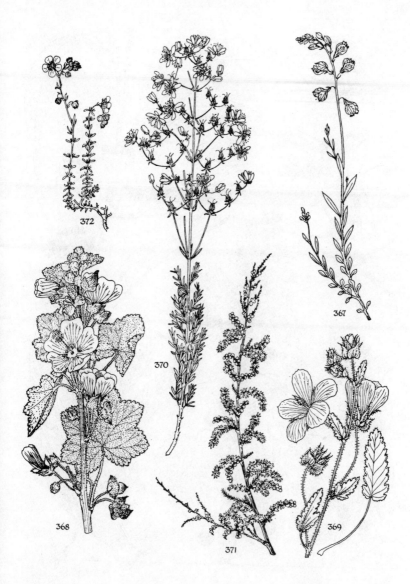

Plate 11 (× ½)
367 Polygala nicaeensis. 368 Lavatera arborea. 369 Malope malacoides. 370 Hypericum empetrifolium. 371 Tamarix gallica.
372 Fumana thymifolia.

Plate 12 ($\times \frac{1}{2}$)
373 Urtica pilulifera. 374 Myrtus communis. 375 Eucalyptus globulus.
376 E. rostrata. 377 Eryngium creticum.

Plate 13 (× ½)
378 Bupleurum semicompositum. 379 Orlaya grandiflora. 380 Fraxinus ornus. 381 Phillyrea angustifolia. 382 Olea europea. 383 Jasminum fruticans.

Plate 14   (×½)
384 Capparis spinosa.   385 Heliotropium europaeum.   386 Cynoglossum columnae.   387 Lycopsis variegata.   388 Alkanna tinctoria.

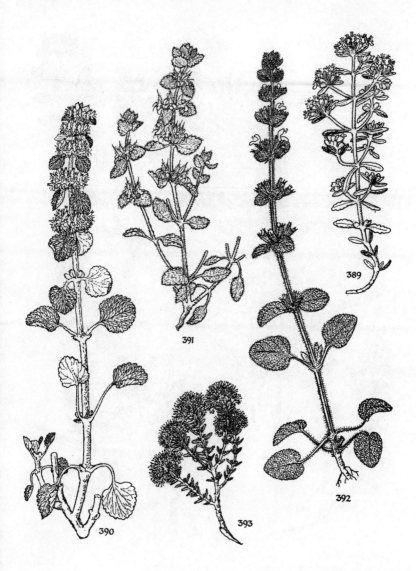

Plate 15   (× ½)
389 Teucrium polium.   390 Marrubium vulgare.   391 Sideritis romana.
        392 Salvia horminum.   393 Coridothymus capitatus.

41

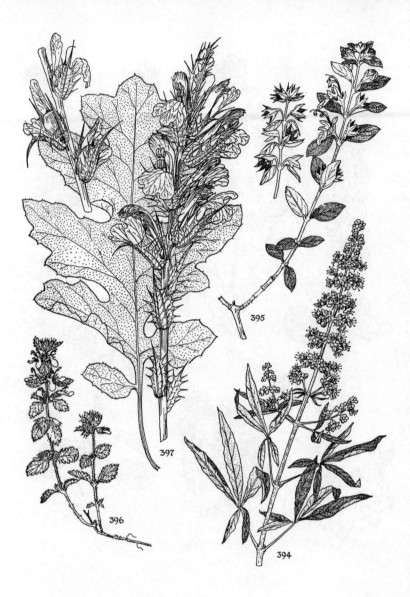

Plate 16 (× ½)
394 Vitex agnus-castus. 395 Teucrium fruticans. 396 T. chamaedrys.
397 Acanthus mollis.

Plate 17   (× ½)
398 Prasium majus. 399 Solanum sodomaeum. 400 Nicotiana glauca.
401 Linaria chalepensis. 402 L. pelisseriana. 403 Veronica cymbalaria.
404 Parentucellia latifolia.

(× ½)

Plate 18
405 Plantago lagopus.   406 Lonicera implexa.   407 Valerianella vesicaria.   408 Scabiosa stellata.

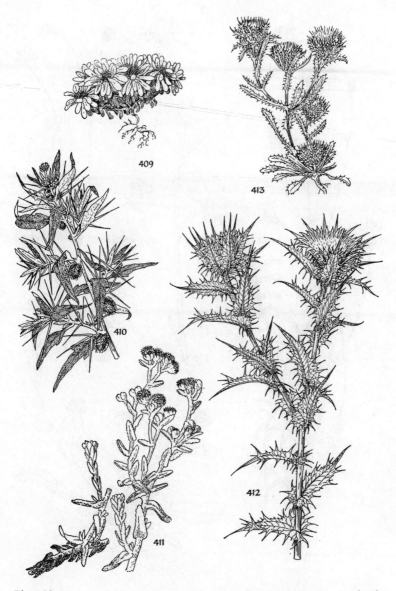

Plate 19 (× ½)
409 Evax pygmaea. 410 Xanthium spinosum. 411 Otanthus maritimus.
412 Carlina corymbosa. 413 Atractylis cancellata.

Plate 20   (× ½)
414 Crupina crupinastrum.   415 Centaurea solstitialis.   416 Rhagadiolus stellatus.   417 Hedypnois rhagadioloides.

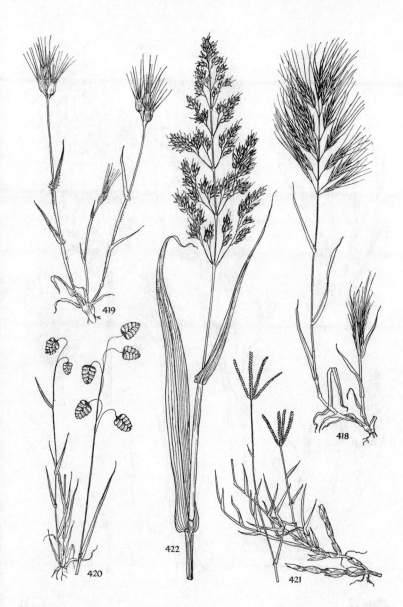

Plate 21 ($\times \frac{1}{2}$)
418 Bromus madritensis. 419 Aegilops ovata. 420 Briza maxima.
421 Cynodon dactylon. 422 Sorghum halepensis.

Plate 22   ($\times \frac{1}{2}$)
423 Hordeum leporinum.   424 Poa bulbosa var. vivipara.   425 Catapodium rigidum.   426 Avena sterilis.   427 Lagurus ovatus.   428 Andropogon hirtus.

Plate 23   ($\times \frac{1}{2}$)
429 Urginea maritima.   430 Hyacinthus orientalis.   431 Asparagus acutifolius.   432 Smilax aspera.   433 Pancratium maritimum.

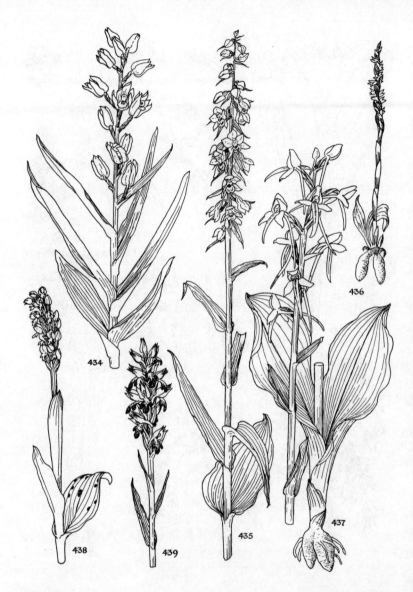

Plate 24 (× ½)
434 Cephalanthera longifolia.   435 Epipactis helleborine.   436 Spiranthes spiralis.   437 Platanthera chlorantha.   438 Neotinea intacta.
439 Orchis coriophora.

50

GNETACEAE—Ephedra Family

This family is probably a living relic of an ancient group of naked-seeded plants. The sexual organs are arranged in small 1-seeded cones which do not become woody as in the pines. The leaves are opposite, there are no resin canals and the wood contains vessels.

EPHEDRA

E. fragilis Desf. ssp. campylopoda (*C. A. Mey.*) *Aschers. & Graebn.*   **312**
JOINT-PINE

A switch-like plant with much-branched, rounded, jointed green stems carrying tiny scale-leaves, which climb through vegetation. Ultimate branches often pendulous, stems up to 4 mm. thick, with narrow rough ribs, very fragile. Cones in pairs; male cones globular of opposite paired scales, flowers of 4–8 stamens; female cones smaller, often on curved stalks, of 1–2 flowers composed of a naked ovule and style. Fruit a bright red fleshy spherical berry, 8–9 mm. in diameter. HABITAT: hedges, bushy places, rocks and walls; Spain, Dalmatia to Palestine, Morocco to Tunisia. February–April.

E. major Host has erect rigid much branched stems, 10–20 cm. high. Joints of stems short, 1–2 cm. long, and stems very thin, about 1 mm. in diameter. Cones solitary. Fruit 5–7 mm. in diameter, scarlet or yellow. HABITAT: rocky places; Spain to Palestine, Morocco to Tunisia. April–June.

The drug ephedrine comes from members of this genus.

CUPRESSACEAE—Cypress Family

A family of evergreen trees and shrubs with small scale-like leaves arranged closely around the branches, or with narrow needle-like and spreading leaves. The cones are 1-sexed. The males are composed of 4–8 whorls of scales bearing clusters of stamens on the upper surface; they are small and soon fall after maturity. Female cones of a few scales with ovules borne on the upper surface, which on ripening become dry and woody, or fleshy and berry-like. Quite an important timber-producing family and commonly planted for ornament.

JUNIPERUS: leaves either spreading and needle-like or in whorls of 3 scales closely pressed to, and covering the branches; female cones of 3–8 scales which become tough and fleshy and fused together in a rounded berry-like fruit.

CUPRESSUS: usually trees with scale-like leaves in whorls of 4 closely pressed to, and covering the branches; cones of 3–6 flat-topped shield-shaped scales pressed together at the margins but becoming woody and separating apart in the second year.

JUNIPERUS

J. communis L. 1   COMMON JUNIPER

A dense bushy grey-green shrub, 1–6 m. high. Leaves needle-pointed, 10–15 mm. long, white on the upper surface with green margins, dark green beneath. Male and female cones yellow, usually on separate plants. Fruit green, berry-like, 6 mm. in diameter, turning bluish-black on ripening during second year. HABITAT: dry calcareous hills, thickets, generally replacing the following species as one ascends the hills; Spain to Turkey, Morocco, Algeria. March–May. British native.

Oil of Juniper is distilled from the green fruits and gives the flavour to gin. A powerful diuretic.

J. oxycedrus L. 313

A silvery grey, very spiny-leaved shrub or small tree which is widespread in the Mediterranean region, 1–8 m. high with dense branches. Leaves awl-shaped, very sharp-pointed, 16 mm. long and distinguished by 2 white bands on the upper surface which are separated by a central green nerve. Fruits 6–10 mm. in diameter, at first green and later turning a shining red-brown at maturity in the second year. HABITAT: rocks, garigue, maquis, woods; circum-Medit. February–May.

Oil of Cade is obtained as a distillate from the wood. The wood is very resistant to decay and hence statues were made of it in classical times; it is also valuable for charcoal making. This may be the biblical 'heath in the desert'.

J. phoenicea L. 2   PHOENICIAN JUNIPER

A shrub or small tree with stout cord-like twigs covered with closely pressed scale-like leaves. Leaves blunt in 4–6 overlapping ranks, scarcely 1 mm, long, with a glandular furrow on the back. Fruit globular, 6–15 mm. in diameter, formed at the ends of the branches, red and shining when ripe in the second year. HABITAT: hills, rocky places on limestone; circum-Medit. February–April.

The wood is very slow to decay; it was used for housebuilding in classical times, and pitch was obtained from this plant. Possibly with the preceding species the 'cedar in the wilderness' of the Old Testament. It is an important forest tree in Morocco.

CUPRESSUS

C. sempervirens L. 314   FUNERAL CYPRESS

A tall tree with rather dark green foliage which either grows as an erect dense columnar tree, or is pyramidal with spreading horizontal branches. Leaves blunt, triangular, scale-like and overlapping, in 4 ranks closely pressed to the stem. Cones globular, 2½–3 cm. long, of 8–14 pentagonal scales, shining brown or greyish; each scale has a sharp-pointed central spine. HABITAT: coastal hills; E. Medit., Greece to Asia Minor, Middle East, N. Africa. April.

Commonly planted throughout the Mediterranean region in gardens and cemeteries, particularly the columnar form, *var. pyramidalis (Targ. & Tozz.) Nym.* which gives such a familiar appearance to Italian landscapes. *Var. horizontalis (Mill.) Gord.*, the spreading form, is less commonly cultivated.

Cypress wood is very hard and durable, and was much used in antiquity for making idols. It was also used in shipbuilding and housebuilding by Phoenicians, Cretans and Greeks. The tree was worshipped in Cyprus and the island is named after it. Biblical 'gopher wood' and 'cypress tree on the mountain of Hermon' refer to this tree.

PINACEAE—Pine Family

A family of tall often conical trees with resinous wood and branches generally in whorls. Leaves needle-like, usually in clusters of 2 or more, or solitary, and ranged along the stems. Male cones small and soon falling, of many spirally arranged scales with anthers on the lower side; female cones of many spirally arranged scales with usually 2 ovules on the upper side. After fertilization the cones swell; they are at first green and later become woody and the scales separate and release the winged seeds. One of the most important timber-producing families giving soft-woods, wood pulp, plywood and also pitch, turpentine and resin.

PINUS: leaves in bundles of 2–3 or 5, needle-shaped and surrounded by chaffy bracts at the base; cones with many scales, first green and becoming woody in the second year.

ABIES: leaves single, arranged spirally along the stem but often appearing flattened and 2-ranked; leaves with a white band, and attached to the stem by a short stalk and a rounded pad which leaves a circular scar when the leaf falls. Cones erect, of many thin overlapping scales which fall away individually from the central axis on ripening.

PINUS

P. halepensis Mill. 3, 315   ALEPPO PINE

A tree with bright green needle-shaped leaves and silvery grey bark, particularly when young. 10–20 m. high, at first branching from the base and later more or less umbrella-shaped. Leaves paired, slender, flexible, 6–10 cm. long and ½–¾ mm. broad. Cones pendulous or with downward-curved stalks, remaining on tree for 3 years before opening; conical, gradually tapering to blunt apex, 8–12 cm. long; cone scales feebly bossed, at first yellowish-brown, darkening to chestnut. HABITAT: limestone hills by the sea, rocky ground; circum-Medit. March–May.

An important and widespread tree which is very drought-resistant and forms open forests extensively in the Mediterranean. An important timber tree of the ancient Greeks, though not as durable as silver fir; used in ship-building, especially for bent wood for triremes according to Theophrastus, also for housebuilding and a source of pitch. Probably the 'fir tree' of the Old Testament. It is tapped for resin and is an important fuel tree.

P. pinea L. **316**   Stone Pine, Umbrella Pine

Mature specimens with their dark green umbrella-shaped crowns are characteristic features of the Mediterranean landscape. Trunk robust, 15–30 m. high, with cracked scaly bark, branched above into dense flat crowns. Leaves paired, rather stiff, 8–15 cm. long. Cones very large, nearly as broad as long, 10–15 cm. by 8–10 cm., reddish-brown, shining, with rhomboidal scales, ripening in the third year. Seeds large and edible. HABITAT: maritime sands and alluvium; Spain to Lebanon. April–May.

Its country of origin is not known but it has been widely planted since Roman times and often forms pure stands in coastal regions. Planted in N. Africa.

P. pinaster Ait. (P. maritima Lam.) 4, 317   Maritime Pine

A tall pyramidal tree, 20–30 m. high, distinguished by its reddish trunk and dark boughs. Leaves thick, stiff and slightly twisted, very long, 10–20 cm. Cones large, 8–18 cm. long, in clusters, bright shining brown when ripe, more or less sharp-pointed. Cone scales with pyramidal shiny 'lacquered' bosses; cones remaining on trees for several years. HABITAT: forms pure forests on acid and siliceous soils; W. Medit., Spain to Italy, Sicily, Morocco to Tunisia. April–May.

Often planted for reclaiming sand-dunes. A valuable source of turpentine and timber.

ABIES—Silver Fir

A. cephalonica Loud. 5   Greek Fir

A pyramidal fir tree with spreading branches and dark green foliage, forming extensive forests in the mountains of Greece. Leaves spreading, narrow, 2–3 mm. broad, linear and ending in a short spiny point, and with 2 white lines on the lower surface. Buds resinous. Cones cylindrical, blunt. HABITAT: mountains above 800 m.; Greece, Yugoslavia. May–June.

The similar *A. alba Mill.* is much more widespread in the mountain regions of Europe. Distinguished by the leaves which are notched at the apex, not sharp-pointed, and by their arrangement into 2 double ranks. Buds not resinous. Cones with bracts, long-pointed and curved at the apex. HABITAT: mountains above 800 m.; Europe. April–May.

The silver fir was described by Theophrastus as the best shipbuilding timber for triremes and long ships because of its lightness and durability; also used for housebuilding.

FAGACEAE—Beech Family

Trees or shrubs with simple or alternately arranged leaves, deciduous or less commonly evergreen; stipules deciduous. Flowers in catkins; male in tassel-like clusters of many-stamened petal-less flowers; female in separate spikes or at the base of the male catkins. Fruit a 1-seeded nut often surrounded by a scaly or spiny cup. A family of many important timber trees such as the oak, ash, beech and sweet chestnut.

QUERCUS

Q. macrolepis Kotschy (Q. aegilops Boiss.)   VALONIA OAK

A deciduous oak with somewhat leathery leaves which fall late in the year. Cups of acorns large, globular, 2–4 cm. in diameter, composed of thick, broadly lance-shaped pointed scales which are widely spreading and often recurved. A short thick-trunked tree, 5–15 m. high, with long-stalked leaves deeply divided into unequal triangular lobes with pointed teeth; upper surface dark shining green, lower with soft grey hairs. Acorns very large, solitary or in clusters of 2 or 3, stalkless. HABITAT: solitary or in open forests in the hills; Italy, Greece to Palestine. April–May.

The cups of this tree are used in tanning and are an important article of commerce; they produce a strong black dye. The acorns are apparently eaten by humans in times of food shortage.

Q. ilex L. 318   HOLM OAK, EVERGREEN OAK

A shrub or small tree, with hairy branches and very variable leaves, which may be spiny. Often mistaken for the Kermes Oak, but distinguished from it by the adult leaves which are white or grey-hairy on the lower surface; and the acorn cup with scales that are all closely pressed to the cup, not spreading. Bark rough; leaves leathery, persisting for 2–3 years, oval-oblong or lance-shaped with entire untoothed margins, or with spiny toothed margins, particularly on young suckering shoots; hairless above, but always hairy below. Male catkins elongated, hairy. Acorn cup hemispherical, grey with woolly hairs and triangular blunt scales; acorn surmounted by a long point. HABITAT: arid places, maquis, woods and hills on limestone; circum-Medit. April–May.

This is the Ilex of the Romans. The wood is hard and resistant; the bark is rich in tannin; it makes good charcoal; the acorns are used for feeding pigs. In classical times the wood was used for houses, shipbuilding and for underground work because it decays slowly.

Q. suber L. 6, 319   CORK OAK

Distinguished by its extremely thick fissured grey bark which, in the Mediterranean, is stripped every 6–12 years leaving the young bark exposed which is at first reddish in colour and later turns grey. A tree, 5–15 m. high, with leathery leaves persisting for 2 years; leaves oval with 10–14 conspicuous secondary veins, and with almost entire or spiny-toothed margins, dull green above and pale grey-woolly beneath. Male catkins elongated, hairy. Acorn cup conical from the base, grey and woolly, with the scales slightly spreading and getting longer from the base of the cup towards the rim. HABITAT: siliceous hills on the littoral; W. Medit., Spain to Italy, Sicily, N. Africa. April–May.

An important bottle cork industry is dependent on this source of cork in the Western Mediterranean.

Q. coccifera L.   320   KERMES OAK, GRAIN TREE

Usually a low shrub with small spiny holly-like leaves, and small acorn cups with stiff pointed spreading scales. ½–3 m. high with very stiff interwoven branches, but if left it may become a small tree. Leaves very leathery, persisting for 2–3 years, oval or oblong, 1–5 cm. long, toothed and spiny (the leaves may be spineless on old branches), bright green and shiny usually on both surfaces, or when young with grey hairs on the lower surface. Male catkins short, hairless. Scales of cup prickly; acorn ovoid or oblong. HABITAT: dry places on limestone and siliceous soils; circum-Medit. March–May.

The Kermes oak is the host plant of the scale insect Coccus ilicis *Planch., the female of which, when dried, gives a beautiful red dye, celebrated from the time of Pliny to Chaucer who writes of the 'Grain of Portugal'; the 'scarlet' of scripture. The bark yields a black dye; it is rich in tannin but because of the small dimensions of the trunk it is seldom used. Theophrastus describes how the wood was used for wheelbarrows, cross-bars of lyres and psalteries, and the stationary pieces of fire sticks.*

MORACEAE—Mulberry Family

An economically important family which is mainly tropical. Trees or shrubs with milky juice, and simple or lobed leaves with 2 stipules which fall off and leave a scar. Flowers in compact heads, 1-sexed; the calyx often becomes fleshy in fruit. The mulberry, fig, bread-fruit and jack-fruit belong to this family.

FICUS

F. carica L.   7   FIG

The fig is widely cultivated in the Mediterranean region, but often occurs in a semi-wild state as a low scrambling shrub, often over rocks, while in cultivation it is a small stiff tree, 2–5 m. high, with smooth grey stems. Leaves large with 3–7 blunt rounded lobes, hairy, rough and deciduous. Flowers 1-sexed enclosed in a fleshy pear-like structure with a small opening in the apex. In cultivation the fig gives 2 kinds of fruit, the first in early summer from figlets of the previous year, and the second in the autumn, of smaller, more numerous and less succulent fruits. HABITAT: in the wild, rocks, woods and garigue; circum–Medit. June–July, August–November.

In cultivation the fig has no male flowers and the fruits ripen parthenogenetically, but in certain parts of the Mediterranean branches of the wild fig, with male flowers, are hung in the fig orchards. A pollinating insect,

56

Blastophaga grossorum, *flies from the wild figs and enters the female cultivated fruits to lay its eggs; pollination is thus brought about and more rapid maturation of the fruits results.*

Figs are of great importance as food for Mediterranean peoples; they are eaten both fresh and dried. The fig was very important in Biblical times; it is referred to more frequently than any other species. Half-ripe figs are considered to be poisonous. The fruit is used medicinally in the east as a remedy against boils and other skin infections; the milky juice of the fresh green fruit is used to destroy warts.

According to Theophrastus fig wood is easily bent and it was thus useful for making theatre seats, hoops, garlands and ornaments.

F. sycomorus L. (**321**) is the 'sycomore' of the Bible and a fig which grows wild in Palestine and Egypt. It has smaller, undivided egg-shaped evergreen leaves and produces fruit several times during the year, on leafless twigs and older stems. It has a light but durable wood from which mummy coffins were made in Egypt.

URTICACEAE—Nettle Family

Herbaceous plants often with stinging hairs on leaves and stems. Flowers small, green, in clusters or spikes at the ends of the stems or in the axils of the simple leaves. Flowers 1-sexed, male with 4–5 sepals and stamens, female with 4–5 sepals and a single 1-seeded ovary. A widespread family including many obnoxious weeds.

URTICA—Nettle

U. pilulifera L. 373   ROMAN NETTLE
A robust stinging nettle with conspicuous spherical fruits, 1 cm. across, on long spreading or pendulous stalks. An annual or biennial up to 1 m. high, with simple or branched erect stems. Leaves oval, rounded at the base with deeply cut margins and long stalks. The whole plant is covered with long bristly stinging hairs. Male flowers on branched axillary stems in interrupted clusters; female flowers in dense rough spherical balls. HABITAT: rubbish dumps, bases of walls and waste disturbed ground; circum-Medit. April–October. Now extinct in Britain; formerly a British alien.

Dioscorides recommends that the juice of this nettle, mixed with oil, be taken to cure sore joints; nettle flagellation has been used as a cure for rheumatism. At Easter in Greece young women beat themselves with this nettle as a reminder of Christ's sufferings.

SANTALACEAE—Sandalwood Family

A small family of herbs, trees or shrubs with undivided leaves. Flowers tiny, often greenish, formed of a calyx of 4–5 often fleshy segments fused to the ovary. Ovary inferior with a single cavity and 1–3 seeds; fruit always 1-seeded.

OSYRIS

O. alba L.   351

A small broom-like shrub with green angular branches, small lance-shaped leaves and tiny yellowish flowers. A rather dense, stiff, erect, branched plant, 40 cm. to 1½ m. high; leaves leathery, evergreen, 1–1½ cm. long. Flowers sweet-scented, 1-sexed, males in small laterally placed clusters, females solitary, borne at the ends of the leafy branches. Fruit fleshy, the size of a pea, bright red. HABITAT: arid, shrubby places; circum-Medit. April–June.

ARISTOLOCHIACEAE—Birthwort Family

A small family of herbs or climbing shrubs largely of the tropical and warm temperate regions. Flowers usually lurid purplish in colour; petals absent: the calyx forms a tube round the fused stamens and style. Ovary inferior, forming a capsule or berry. Leaves heart-shaped, stalked and entire-margined. Some species are used as a cure for snake bites.

ARISTOLOCHIA—Birthwort, Pipe Vine

A. sempervirens L. (A. altissima Desf.)   8

A climbing plant with egg-shaped leaves, and solitary dull purple, curved euphonium-like flowers. Leaves evergreen, pointed with heart-shaped base, 4–8 cm. long by 5 cm. broad. Flowers U-shaped and funnel-like with an oblique lip which is shorter than the basal part of the tube; base of tube swollen. HABITAT: hedges and shady places; E. Medit., Sicily, Italy to Palestine, N. Africa. January–June.

A. rotunda L.   9

An unusual looking plant with neat rounded leaves and small brown straight funnel-like flowers with a long tongue-like lip bent over the opening.

The swollen tuberous root produces herbaceous stems, 20–60 cm. high. Leaves stalkless, 4–6 cm. broad, oval, blunt with smooth margins and heart-shaped lobes nearly touching or overlapping. Flowers with a straight yellow tube and a brown lip as long as, or longer than the tube. HABITAT: fields, hedges and stony places; circum-Medit., not Morocco. April–June.

The similar *A. pistolochia L.* has roots formed of a tuft of narrow elongated fibres. The plant is covered with short stiff hairs. Leaves smaller, 2–3 cm. broad, with a rough toothed margin and with the basal lobes of the blades widely spaced. Flowers with a lip shorter than the tube of the calyx, brownish in colour. HABITAT: dry and arid places; W. Medit. to France, Corsica, Sardinia, Morocco.

RAFFLESIACEAE—Rafflesia Family

Fleshy parasites which grow on the roots and stems of plants, and possessing scale-like leaves which are never green. Flowers large and solitary (Rafflesia has the largest known flower in the plant kingdom), or in heads. Fruit fleshy, many-seeded. A small family of tropical or temperate parasitic plants.

CYTINUS

C. hypocistis (*L.*) *L.* **10**

A most extraordinary parasitic plant growing on the roots of cistus species. It has a globular fleshy head of bright lemon yellow flowers encircled by scales which are often bright carmine. The flower-heads push through the soil below cistus bushes; there are no green leaves. Stems 4–8 cm. high, closely covered in brightly coloured oval overlapping scales. Flowers 5–10, in a rounded terminal head, calyx tubular, 4-lobed; the upper flowers male, the lower female. Fruit soft, pulpy. There are several sub-species with different colouring of flowers and bracts, which parasitize different cistus species. HABITAT: maquis and garigue; circum-Medit. April–June.

AMARANTHACEAE—Amaranth Family

Coarse annual plants with sometimes showy plume-like clusters of tiny purplish or greenish flowers. Leaves stalked and undivided. Petals absent, sepals and bracts subtending flowers dry and membraneous; stamens 3–5; ovary with 2–3 stigmas, usually 1-celled and opening by a lid. Many are weeds of cultivation and a few, such as Love-lies-bleeding, are grown ornamentally.

AMARANTHUS

A. retroflexus L. **322**   PIGWEED

A robust hairy annual with pale green, broadly lance-shaped leaves and a terminal plume of numerous tiny greenish flowers. An erect slightly branched plant, 20–80 cm. high, with the stem covered with many short rough hairs. Leaves up to 15 cm. long; narrowed into a rough stalk. Flowering heads short, dense, greenish-white, the upper part leafless. Bracts round flowers stiff, somewhat prickly, 3–4 mm. long, twice as long as the 5 sepals; stamens 5; fruit ovoid, about 1½ mm. long and longer than sepals. HABITAT: cultivated ground, waste places, track sides; circum-Medit. July–September. A native of Tropical America. British casual.

The generic name is sometimes spelt *Amarantus* by Continental authorities.

NYCTAGINACEAE—Four o'clock Family

A family of trees, shrubs and herbs with simple leaves without stipules. Flowers usually surrounded by a group of bracts (involucre) which may be brightly coloured. Petals none, calyx often petal-like and bell or trumpet-shaped. Fruit a 1-seeded winged or grooved carpel, often invested by the calyx. Some species are grown for ornament.

BOUGAINVILLEA

B. spectabilis Willd. **11**

A Brazilian plant introduced to Europe in 1829 and now very widely planted in villa gardens for covering walls and buildings, and for giving shade. A tall spiny climbing plant with small blunt oval leaves in whorls. It has brilliantly coloured floral bracts of varying shades of red, violet, purple or orange (the flowers themselves are insignificant), and these bracts appear at the end of February and last throughout the summer; in North Africa they appear all the year round. Not naturalized.

AIZOACEAE—Mesembryanthemum Family

A family of mainly succulent plants which are characteristic of semi-desert regions, sea shores and salt-rich soils. Herbs or low shrubs with swollen mostly undivided opposite leaves. Flowers solitary, often conspicuous, with calyx tube fused to the ovary; petals and stamens very many in our species, the whole flower recalling a member of the daisy family. Fruit a capsule or nut-like. Many species grown for ornament.

MESEMBRYANTHEMUM: annuals or biennials; leaves covered with small projections.

CARPOBROTUS: perennials with thick woody 2-angled branches, and large, smooth, 3-angled curved opposite leaves united at the base. Flowers very large.

LAMPRANTHUS: similar to Carpobrotus but smaller, stems not angled, with cylindrical or 3-angled leaves.

MESEMBRYANTHEMUM

M. nodiflorum L.

A spreading, branched, fleshy-leaved plant with curious crystalline shining hairs on its stems giving the appearance of hoar frost, and daisy-like flowers with many narrow white petals. An annual up to 30 cm. high with only the upper branches covered with crystalline hairs. Leaves linear, fleshy, cylindrical, 1–2½ cm. by 1–2 mm. Flowers small, solitary, stalkless, in the axils of leaves; petals white, yellow at the base, shorter than the calyx. HABITAT: salt marshes, sands and rocks by the sea; circum-Medit. April–July.

The similar *M. crystallinum* L. has fleshy flattened oval leaves, 7–12 cm. long, and densely covered, like the rest of the plant, with swollen transparent crystalline hairs. Flowers larger, 2–3 cm. across, petals longer than calyx. Fruit red and succulent, the size of a nut. HABITAT: salt marshes, sands and rocks by the sea; circum-Medit. (not Turkey). April–June.

CARPOBROTUS

C. edulis (L.) N.E. Br. **13**   HOTTENTOT FIG

A creeping plant forming extensive mats of large triangular-sectioned, upwardly curving, paired leaves, with solitary, brilliant pale lilac, yellow or orange flowers with many petals. Stems trailing, woody, rooting at nodes; leaves 7–10 cm. long, fused in pairs by their bases. Flowers borne on swollen stalks, 8–10 cm. across, with numerous linear spreading petals, numerous stamens and 5-lobed calyx fused to the ovary. Fruit fleshy and edible. HABITAT: A plant from the Cape which has become widely naturalized in the Mediterranean region on rocks, cliffs and sands by the sea. April–July. Naturalized, particularly in the West, in Britain.

The similar *C. acinaciformis (L.) L. Bol.* **(12)** is more robust and is distinguished by its startlingly bright red-carmine flowers which are 12 cm. across, and the leaves which, when viewed from the side, are broader towards the apex than the base, and narrowly triangular in section. Fruit fleshy, edible. HABITAT: A plant from the Cape naturalized in similar localities to the last. April–July.

LAMPRANTHUS

Lampranthus roseus (Willd.) Schwant. **14**

An erect or sprawling, woody-stemmed plant with narrowly triangular-sectioned leaves and large pink flowers. Leaves fleshy, 2½–3 cm. long, up to 4 mm. wide, covered in transparent dots. Flowers on 5 cm. stems, 4 cm. across, with numerous narrow petals and stamens. HABITAT: A plant from the Cape which is widely planted in the Mediterranean and may become naturalized by the sea. Spring and summer.

There are many other species with pink, carmine or yellow flowers similarly used for decorative purposes.

CARYOPHYLLACEAE—Pink Family

A family of herbaceous annuals and perennials with pairs of narrow, undivided and opposite leaves which are often fused together at the base and encircle the stem. Flowers usually in a loose dichotomously branching head, rarely solitary. Sepals 4–5, free or united into a tube; petals usually 4–5, unfused and often deeply divided into 2 lobes; ovary often shortly stalked, 1-celled, with the ovules arranged on a central boss (free central placentation); styles 2–5 free or partly fused. Fruit a dry capsule splitting by teeth or valves. The Mediterranean is the main centre of distribution of this family which has many ornamental garden plants including the carnation.

SILENE: sepals fused into a tube with 5 triangular teeth and often with 10 conspicuous veins (or multiples of 10); styles 3–5, teeth of capsule double this number.

61

AGROSTEMMA: flowers conspicuous, solitary or few; sepals fused into a 10-ribbed tube with 5 long teeth, longer than the petals; styles 5.

DIANTHUS: leaves paired from swollen joints, narrow-linear and bluish grey; flowers with additional sepal-like segments (epicalyx) of 1-3 pairs below each flower; sepals joined into a smooth cylindrical tube; petals pink or white; styles 2.

VACCARIA: sepals fused into a tube with 5 sharp angles or wings and 5 triangular teeth; styles 2.

SAPONARIA: sepals fused into a green tube which is smooth and un-winged; petals with a scale at the throat of corolla; styles 2-3.

KOHLRAUSCHIA: flowers in oval heads surrounded by loose bracts with papery margins; sepals joined below into a 15-veined tube with thin whitish seams; styles 2.

PARONYCHIA: leaves opposite or alternate and furnished with papery stipules; flowers very small, subtended by silvery bracts; sepals unfused; ovary with 2 styles, 1-seeded and unsplitting in fruit.

SILENE—Campion, Catchfly

S. vulgaris (Moench) Garcke (S. cucubalus Wib., S. inflata Sm.)
BLADDER CAMPION

A white flowered, greyish leaved, usually hairless plant easily distinguished by its blown-up, balloon-like calyx which has 20 rather conspicuous branching veins over its surface. A perennial, branched from the base, up to 90 cm. high, with oval to lance-shaped leaves, the lowest stalked, the upper ones half clasping the stem. Flowers in spreading terminal clusters, each flower nearly 2 cm. across, with deeply bilobed petals; calyx egg-shaped hairless, enlarging in fruit, with 5 broad triangular teeth. A very variable plant. HABITAT: cultivated ground, sands and rocks; circum-Medit. April–August. Native in Britain.

S. gallica forma quinquevulnera (L.) Mert. & Koch 323

This form of the SMALL-FLOWERED CATCHFLY is a distinctive plant with pink petals each with a large dark crimson spot at the base. A hairy glandular erect, branched annual, 15-40 cm. high, with lower leaves obovate and upper linear acute. Flowers 10-12 mm. across, arranged along one side of the elongated flower stems; calyx sticky, ovoid in fruit, 10-veined, with long spreading hairs and 5 long-pointed teeth; stamens with hairy stalks. The typical *S. gallica L.* is widespread and has small white or pinkish flowers; it is native in Britain; the form is native in the Channel Is. HABITAT: cultivated ground, track sides; circum-Medit. March–June.

S. succulenta Forsk. 15

A succulent plant with hairy, velvety and sticky leaves to which sand-grains may adhere, and rather large white typical campion-like flowers.

A perennial, woody at the base, with many leafy stems, forming low-growing clumps up to 20 cm. high. Leaves oblong, 2–4 cm. long, broader towards apex, blunt and closely crowded on the stems. Flowers nearly 2 cm. across with deeply divided petals, short-stalked, forming a leafy cluster at the ends of the stems; calyx 2 cm. long, hairy and sticky like the leaves. HABITAT: maritime sands; N. Africa, Corsica, Sardinia, Crete, Asia Minor to Palestine. April–June.

S. colorata Poir. 16
A small rough-hairy annual with rather 1-sided heads of bright pink flowers and a cylindrical calyx with 10 reddish unbranched veins. Branched from the base, 10–30 cm. high, without sterile basal shoots. Leaves 1–3 cm. long, hairy, broader towards apex and narrowed into long stalks; upper lance-shaped. Flowers in loose clusters of 1–4; petals bilobed, longer than calyx; calyx cylindrical and shortly hairy, with blunt triangular teeth, becoming club-shaped in fruit. Seeds with winged undulating margins. HABITAT: cultivated ground, stony places and sand by the sea, often in abundance; circum-Medit. (not France). April–June.

The similar *S. aegyptiaca L.* is probably the commonest annual campion in the Asiatic Mediterranean, and in winter it carpets ploughed fields and olive groves. It has a loose head of a few conspicuous bright rosy purple flowers, 2 cm. across, with red calyx. It is distinguished by its stalked egg-shaped leaves, its petals with a tooth on either side of the base, and by the reddish calyx which is faintly veined, cylindrical from a swollen base and with blunt egg-shaped teeth with membraneous margins. HABITAT: fields, olive groves; Turkey to Palestine, Egypt. January–April.

S. behen L. 324
An erect hairless and somewhat greyish-green annual with a dichotomously branched head of pink flowers and an ovoid calyx with conspicuous branched reddish veins. Up to 40 cm. high with paired hairless leaves, the lower stalked, obovate to lance-shaped, the upper stalkless. Flowers short-stalked, pink; calyx hairless, 10–11 mm. long, with 10 veins which are branched and form a network in the upper constricted part, and with a swollen base considerably broader than the flower stalk. Fruit included within calyx; seeds rough, granular, not smooth. HABITAT: fields, hills, rocky places; Sardinia, Italy to Palestine, N. Africa. March–May.

AGROSTEMMA

A. githago L.   CORN COCKLE
An annual with large spreading reddish-purple flowers borne at the ends of long stems, and with 5 narrow pointed sepals spreading well beyond the petals. An erect and sparingly branched annual with spreading white hairs, and leaves in opposite pairs, narrow lance-shaped, hairy, 5–12 cm. long. Flowers 3–5 cm. across, solitary and terminal; petals broad, rounded and slightly notched, veined with deep purple; calyx hairy,

10-ribbed with long narrow pointed sepals. Fruit ovoid, longer than calyx tube. HABITAT: cornfields; circum-Medit. April–June. Naturalized in Britain but disappearing.

Often a troublesome weed in cornfields; the seeds are poisonous and damage the physical properties of wheat flour. It is the 'cockle' of the Bible. In Roman times, guests attending feasts and games were presented with a coronet made of these flowers.

DIANTHUS—Pink

D. strictus Banks & Sol. (D. multipunctatus Ser.)
The commonest pink to be found in the Eastern Mediterranean. The pink flowers have toothed petals and are generally dotted with crimson spots. A perennial with many stems arising from the base, 30–50 cm. high. Leaves rather flabby, narrow lance-shaped with 3 veins and short sheaths. Flowers solitary, 12–20 mm. across; calyx half the length of the petals and characteristically minutely warted and pitted over the surface. Bracts below calyx 4 with membraneous margins and a short terminal spine. HABITAT: dry rocky places; Crete, Turkey to Palestine, Cyprus, Egypt. May–July.

VACCARIA

V. pyramidata Med. (Saponaria vaccaria L.) **325**
An annual with greyish-green narrow paired leaves and a spreading head of small pink flowers with conspicuous angular green-ribbed calyx. Plant hairless, up to 60 cm. high, regularly branched above with many flowers on long stalks. Leaves 1-nerved, lance-shaped, acute, the upper heart-shaped at the base and clasping the stem. Flowers about 8 mm. across; petals toothed, without scales in the throat; calyx ovoid, narrowed at apex with triangular pointed teeth and with 5 pronounced green-winged ribs. HABITAT: among corn, cultivated ground, fields on limestone and clay; circum-Medit. March–June. A British casual.

SAPONARIA—Soapwort

S. calabrica Guss. (S. graeca Boiss.) **17**
A low spreading slender-stemmed annual with red nodes and rose-coloured flowers. Stems much branched at the ends, more or less glandular-hairy; leaves 1-veined, lower narrow-oblong, broader towards the tip, upper oblong lance-shaped. Flower stalks shorter than calyx; petals with almost circular blades and with 2-lobed scales at the throat; calyx 8–10 mm. long, hardly swelling in fruit. HABITAT: rocky places; Italy, Greece, Crete. April–June.

KOHLRAUSCHIA

K. velutina (Guss.) Rchb. (Dianthus velutinus Guss., Tunica velutina Fisch. & Mey.) **326**
A slender erect annual with narrow grass-like leaves and an egg-shaped terminal head of small bright pink flowers closely enfolded in several pairs

of broad shining membraneous bracts. Plant 20–50 cm. high, branched only from the base, glandular-hairy below, hairless above, with few pairs of leaves. Leaves linear, pointed, 1–3 cm. long, with sheaths 2–3 times longer than broad encircling the stem. Bracts surrounding flower head egg-shaped, thin, dry and flexible, the outer terminating in a fine point. Flowers opening one at a time, small, about 8 mm. across, with lobed petals. HABITAT: fields, sandy and rocky places; circum-Medit. March–April.

The similar *K. glumacea* (18) (*Chaub. & Bory*) *Hay.* lacks the points to the bracts surrounding the flower heads, and the petals are somewhat larger, usually with rounded teeth, and purple not rose coloured. HABITAT: grassy hills, rocks; Yugoslavia, Greece, Crete. May–June.

The similar *K. prolifera* (*L.*) *Kunth* has hairless non-glandular stems and the sheath of the leaves is as broad as long. HABITAT: dry, rocky places, track sides; European Medit., N. Africa. March–June. Rare British native.

PARONYCHIA

P. argentea Lam. 327

A low spreading plant with paired lance-shaped leaves and rounded flower heads which are covered in silvery 'tissue paper' bracts. A perennial with branches spreading over the ground to 20–30 cm. Leaves 1–1½ cm. long with short stiff hairs along the margins; stipules papery and shorter than leaves. Flowers in rounded heads towards the ends of the branches, almost entirely covered by the silvery oval bracts which are much longer than the flowers. Sepals 5, oblong, arched and papery, with a fine point; stamens 5. HABITAT: dry rocky places; circum-Medit. April–June.

The similar *P. capitata* (*L.*) *Lam.* is a mat-forming plant with very conspicuous globular silvery flower heads which tend to cover the whole plant after flowering. Bracts silvery, broadly orbicular, much longer than the flowers. Sepals unequal, the 3 outer noticeably longer and with recurved points. HABITAT: dry rocky places; circum-Medit. March–June.

P. echinata (*Desf.*) *Lam.* has much smaller flower heads in the axils of the leaves, and the heads have no conspicuous papery bracts. Heads not overlapping but widely spaced up the stem and shorter than the leaves. Sepals green, becoming dark, with long-pointed and recurved almost spiny tips, and with narrow membraneous margins. A more spreading erect plant than the last with leaves with rough-toothed margins. HABITAT: dry places by the sea; circum-Medit. March–May.

RANUNCULACEAE—Buttercup Family

A family mostly of herbaceous plants (Clematis is generally woody) with usually dissected compound leaves alternately arranged, and conspicuous flowers which are often few or solitary. Flowers bisexual with sepals and petals often similar or with one whorl missing; stamens numerous. Ovary often of many carpels or less commonly a few many-seeded pods. Many botanists consider this to be one of the most primitive of families. Many members of the family are acrid and poisonous to both man and stock; many are grown for ornament.

PAEONIA: Flowers regular, very large; red, pink or white; carried singly. Fruits of large carpels. (Identification can usually be confirmed by the country the species is found in.)

HELLEBORUS: flowers usually greenish with tubular nectaries; fruits of 3–8 pods; leaves large, divided into finger-like segments which are toothed.

NIGELLA: flowers often showy, blue, white or yellowish, the sepals petal-like, the true petals insignificant; fruits unusual in having the pods either partially or completely fused along their inner margins; leaves dissected into fine linear segments.

DELPHINIUM: usually perennials with long spikes of blue flowers; flowers symmetrical in one plane only, with a backward-projecting spur; fruits of 3–5 splitting pods.

CONSOLIDA: annuals with spurred flowers but with a fruit of a single many-seeded splitting pod.

ANEMONE: flowers usually solitary and conspicuous with 3 leafy bracts, or sepal-like bracts, below the flower; flower segments coloured (not true petals, but described as such here), many, up to 20; fruits many of 1-seeded units.

CLEMATIS: woody climbers with opposite leaves, and which climb by means of their leaf stalks; petals usually 4, and fruits with long feathery styles.

RANUNCULUS: flowers usually bright yellow in spreading heads, rarely white or red; sepals green; petals with a nectar-secreting flap at the base; fruits of many 1-seeded units.

ADONIS: flowers red or yellow, usually solitary; petals 3–20, with no nectary; fruit an elongated head of many 1-seeded units which are wrinkled and shortly beaked; leaves much divided into linear segments.

PAEONIA

P. mascula (L.) Mill. (P. corallina Retz.)   PEONY

A large-leaved many-stemmed herbaceous perennial with very large solitary reddish-purple flowers. Underground stem swollen, aerial stems leafy, up to 1 m. high. Leaves twice divided into 3's with 9–13 untoothed acute egg-shaped leaflets, hairless on the under surface. Flowers very large, up to 10 cm. across; rarely yellow (Sicily); stamens numerous with crimson stalks and yellow anthers, sepals 5 or more. Fruit covered with shaggy hairs. HABITAT: bushy places and meadows in the hills; Algeria, Trieste, Cyprus, Sicily and Syria. March–May. Naturalized in Britain on Steep Holm.

This plant was used medicinally by medieval monks, hence its occurrence in Cyprus, Syria and Britain. Often cultivated.

66

P. arietina Anderson occurs from N. Italy to Asia Minor, though probably extinct in Greece. It closely resembles *P. mascula* but has 15–17 wider, larger leaflets.

P. russi Bivona is local to Corsica, Sardinia and Sicily.

P. officinalis L. reaches 35 cm. and is distinguished by the leaflets which are further divided into 20 or more narrow lance-shaped segments which become pale bluish green underneath at maturity. HABITAT: bushy places, meadows; S. France, Dalmatia, Albania. May–June.
Often cultivated.

P. humilis Retz. occurs on the Costa Brava, Spain, and in S.W. France. 25–40 cm. high with deeply cut leaves of 30 or more oblong-elliptic segments. The flowers are red and have 2–3 hairless carpels.

P. coriacea Boiss. is a species from Morocco, Algeria and S. Spain with 9–14 dull green broadly elliptic leaflets. About 50 cm. high, with rose-pink flowers giving rise to 2 hairless carpels.

P. broteroi Boiss. & Reut. (28) is more common in S. Spain. Up to 35 cm. high. The leaves are shining, reddish, cut into about 20 narrow elliptic leaflets. Flower carmine or purplish with 2–4 hairy carpels.

P. rhodia Stearn is a white-flowered species endemic to Rhodes.

HELLEBORUS—Hellebore

H. cyclophyllus Boiss. **19**
A rather robust perennial with leafy heads of a few large open green flowers, and conspicuous hand-shaped leaves. Leaves mostly from the rootstock, appearing in winter or early spring, of 5–9 lance-shaped toothed and sometimes lobed segments. Flowers 5 cm. across, 3–4 on a stem up to 40 cm. high, which is leafy to the apex with blades divided into 3–5 segments, hairy below. Fruit of 5 carpels, not fused together. HABITAT: woods and bushy places in the hills and mountains; Greece. February–March.

H. foetidus L., BEAR'S-FOOT, STINKING HELLEBORE, has similar green flowers which are bell-shaped or globular at maturity, not spreading, usually with a reddish-purple margin at the base of the 'petals'. Flowers 1–3 cm. across, numerous, drooping. Root leaves absent, lower stem leaves dark, upper paler coloured, hand-shaped. A strongly smelling plant when bruised. HABITAT: bushy and rocky places, track sides; Spain to Italy. January–March. British native.

NIGELLA—Love-in-a-Mist

N. arvensis L. **328**
An annual with single rather conspicuous pale blue flowers borne at the ends of branched stems, and with finely divided feathery foliage. Stem 10–30 cm. high with long spreading branches; leaves twice cut into narrow acute segments. Flowers about 3 cm. across, often netted with green veins

on the outside; carpels 3–5, with coiled styles when young, and fused together along their inner margins to about the middle, and each 3-nerved. HABITAT: fields and dry hills on calcareous soils; circum-Medit. June–August.

The similar *N. hispanica L.*, the FENNEL FLOWER, has larger flowers, 4–5 cm. across, of a bright blue, with prominent reddish purple stamens. An erect plant 20–40 cm. high, with erect side branches. Sepals with no 'claw' at base. Fruit very glandular. HABITAT: fields, waste places; S. Spain, naturalized in S. France. Summer flowering.

The similar *N. damascena L.* is distinguished from the latter two by the leafy much dissected bracts which surround each flower closely. In habit it resembles *N. hispanica*. Flowers bright blue; fruit of 5 carpels fused along their whole length, and becoming inflated and globular. HABITAT: fields and rocky places; Spain to Turkey, N. Africa. May–June.

This is the commonly cultivated Love-in-a-Mist.

N. ciliaris DC. is a yellowish-white-flowered plant found in cornfields in the E. Mediterranean. Petals, sepals and carpels covered in coarse stiff hairs. Carpels 5–15, fused to about a third of their length, prominently 3–nerved. HABITAT: cultivated ground; Cyprus, Turkey to Palestine. April–June.

DELPHINIUM

D. peregrinum L. 329   VIOLET LARKSPUR

An annual with spikes of blue-violet flowers and almost straight spurs, and upper stem leaves few, narrow and undivided. Stems finely hairy, with a whitish bloom, 20–50 cm. high, with stiff straight almost leafless branches. Lower leaves much dissected into narrow segments. Flowers with unfused petals and slightly upturned spur, about 1½–2 times as long as the rest of the flower. A very variable species. HABITAT: fields and dry stony places; Central & Eastern Medit. (not Italy). May–June.

The similar *D. halteratum S. & S.* (20) is a lower-growing annual with fruits of 3–5 hairy carpels. Stems 15–35 cm. high densely covered with crisp hairs. Flowers bright blue with spurs longer than coloured sepals. Spikes short spreading; lower petals without beard-like hairs. HABITAT: fields, track sides, arid places; circum-Medit. May–August.

D. staphisagria L. has a long hairy spike of blue flowers with very short sac-like spurs only 3–4 mm. long, hardly a quarter the length of the petals. Stems up to 1½ m. high; leaves of 5–9 broad lobes. Carpels 2–4 swollen, hairy. HABITAT: field verges, garigue; circum-Medit. May–August.

A very poisonous plant; used for neuralgia and as an insecticide.

CONSOLIDA

C. ambigua (L.) Ball & Heywood (Delphinium ajacis L.)   LARKSPUR

An attractive annual with rather loose spikes of bright blue, rarely pink or white flowers, and much divided rather feathery leaves. A hairy, sparingly branched annual, 25–60 cm. high, with lower stem leaves long stalked, deeply cut into narrow linear segments, and upper leaves stalkless. Flowers in clusters of 4–16, about 1½ cm. long with a spur about 1½ cm. long; flower stalks hairy. Fruit a single hairy carpel, 1½–2½ cm. long, gradually narrowing into a beak; seeds black. HABITAT: fields and dry places; circum-Medit. April–July. Casual in Britain. Commonly cultivated and sometimes escapes.

Larkspurs were used to garland mummies in Egypt, and after 3,000 years the blue colour of the flowers shows little fading.

ANEMONE

A. blanda Schott & Kotschy   **21**
A small blue-flowered anemone, similar in general appearance to the Wood Anemone. A perennial with a tuberous rootstock and basal leaves 2–3 times dissected; leaves on flower stem 3 times dissected, hairless below and often flushed with violet. Flowers solitary, blue to almost white, sometimes bicoloured, with 9–15 narrow petals which are hairless on the outside. HABITAT: bushy places and rocky ground in the mountains; Greece to Lebanon. March–April.

A. coronaria L.   **24, 26, 27** (Fig. Id, f)   POPPY or CROWN ANEMONE
This is one of the most memorable and beautiful Mediterranean plants because of its large brightly coloured flowers which are among the first to appear in early spring. Leaves all basal, long-stalked, 3 times cut into linear segments and arising from a brown cylindrical tuber. Stem 20–40 cm. high, bearing a solitary flower. Bracts close below flower leaf-like, twice

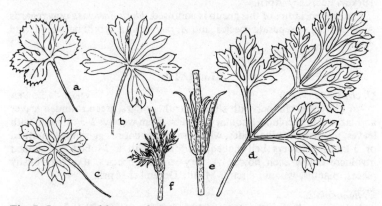

Fig. I. Leaves and bracts of *Anemone* spp. (all × ½)
a Leaf of *A. hortensis* var. *pavonina*   **b** Leaf of *A. hortensis* var. *stellata*
c Leaf of *A. hortensis* var. *stellata*   **d** Leaf of *A. coronaria*   **e** Bracts of *A. hortensis* var. *stellata*   **f** Bracts of *A. coronaria*

cut into narrow segments. Flowers large, 4–8 cm. across, with 5–8 oval petals, in many colour forms of lavender, blue, purple, red, rose and rarely white, often with a differently coloured centre. The red form *coccinea* and the purple form *cyanea* are the most widespread. Fruiting heads in an oval or cylindrical spike and fruitlets covered with white silky hairs. A very variable species. HABITAT: fields, olive groves, vineyards, etc.; Spain to Palestine, Egypt, Algeria. January–April.

Many people regard this plant as the lily of the field which 'surpassed Solomon in all his glory', but several species have equal claim and may well be referred to. Commonly cultivated as St Brigid and de Caen Anemones, etc.

A. hortensis L.   22, 23, 25 (Fig. Ia, b, c, e)

Rather similar to the preceding species with many colour forms but the petals are narrower, more numerous, and the leafy bracts, which are far less finely dissected, are lance-shaped and lie some way below the flowers. A perennial, 10–20 cm. high, with leaves showing considerable variation; the outer basal leaves are often cut into broad rounded lobes, and the inner into narrow lance-shaped segments. Bracts below flowers 3, uncut or with 3 lobes at the apex. Petals 7–18, narrow and usually pointed. It is a very variable species and the flowers may be red, purple, rose and rarely white. The commonest forms are *var. stellata* (*Lam.*) G. & G. (22) with 9–14 elliptic lance-shaped petals, pink-mauve, rarely white, with a more westerly distribution from France to Greece and Crete; *var. pavonina* (*Lam.*) G. & G., with 7–9 broad oblong-ovate or ovate lance-shaped petals, commonly with a white base. One form of this with scarlet flowers (*subvar. fulgens*) (25), is a more eastern plant of Greece and Turkey. HABITAT: cultivated ground and abandoned cultivations, olive groves; France to Turkey. February–April.

The nomenclature of this group is confused. *Flora Europaea* now regards *A. pavonina* as a separate species, and *A. fulgens* as a hybrid between it and *A. hortensis*.

CLEMATIS

C. cirrhosa L.   330   VIRGIN'S BOWER

An evergreen climber with shining leathery dark green rounded leaves and large cream-coloured solitary flowers. Stem woody, 2–5 m. long, with leaves, in clusters at the nodes, which are either entire egg-shaped, toothed or 3-lobed. Flowers bell-shaped, nodding, with a 2-lobed cup-shaped involucre below each flower. A very variable species. HABITAT: bushy places, maquis, woods; circum-Medit. December–April.

C. flammula L.

A climbing plant similar to our Old Man's Beard but with twice or thrice dissected leaves with somewhat thick oval or linear leaflets, often greyish in colour. Flowers 12–30 mm. across in loose heads, very sweet-scented; petals 4–5, white, lacking hairs on the inner surface. Fruits with short feathery styles. HABITAT: hedges, thickets and waste places; circum-Medit. May–August. Sometimes escaped from gardens and locally established in Britain.

RANUNCULUS

***R. ficaria** L. (Ficaria verna Huds.)* **29**   LESSER CELANDINE
A slightly fleshy plant with bright shining yellow flowers and rather glossy heart-shaped leaves. Stem 5–30 cm. high, sparsely branched, bearing terminal flowers and stalked leaves with sheathing bases and blades often shallowly lobed or angled. Flowers 2–3 cm. across, with 6–12 petals, twice as long as the 3 green or yellowish sepals. *Var. grandiflora Rob.* **(29)** is a more robust plant with leaves up to 5 cm. broad and flowers 3–5 cm. across. The sepals are whitish yellow with papery margins. It is the commonest form in the Mediterranean region. HABITAT: damp, shady places, banks, woods and thickets; circum-Medit. February–April. British native.

***R. asiaticus** L.* **30**   TURBAN BUTTERCUP, SCARLET CROWFOOT
The most brilliant and striking of all buttercups with large bright flowers which are often scarlet but may be white, yellow or orange, and with much-elongated fruiting heads. A perennial with a sparingly branched stem up to 30 cm. high bearing 1–5 large terminal flowers. Leaves very variable – root leaves either undivided or cut into wedge-shaped segments, stem leaves more narrowly dissected. Flowers large, 3–6 cm. across; long-stalked and with green spreading sepals; the scarlet form is common in Rhodes, Asia Minor and the Levant, the white form common in Crete, and the yellow-orange forms local in Crete, Cyprus, Asia Minor and North Africa. Stamens numerous, blackish-violet; fruiting heads cylindrical. HABITAT: fields, bushy places; Crete, Turkey to Palestine, N. Africa. February-May.
This buttercup can be confused with the red and white forms of *Anemone coronaria*, but it is distinguished by the presence of sepals and by the leaves which are much less divided.

***R. orientalis** L. (R. myriophyllus DC., R. millefc' ·s Banks & Sol.)* **331**
A small spreading perennial with small pale yellow flowers and leaves which are several times divided into narrow pointed segments and usually densely hairy. Stem 5–15 cm. high, with many spreading branches ending in long flower stalks which sometimes swell in fruit. Flowers 1–1½ cm. across, yellow or whitish, with sepals reflexed at time of flowering; carpels smooth and ending in a long narrow curved beak, hooked at the end. HABITAT: fields and sandy places by the sea; Greece to Palestine, N. Africa. March–April.
A closely related plant is ***R. isthmicus** Boiss.* with 1–4 flowers and fruits with shorter triangular beak, half the length of the fruit. It occurs in Sicily, Greece and the Cyclades.

***R. muricatus** L.* **332**
A nearly hairless annual with inconspicuous pale yellow flowers,

71

distinctive rounded leaves and fruits with conspicuous spines. Stem spreading, 10–30 cm. high, with long-stalked leaves with circular or kidney-shaped blades and shallow rounded-lobed margins. Flowers 8–10 mm. across, petals little longer than the reflexed sepals. Carpels 6–12, about 8 mm. long, flattened and with spines on each face and a distinctive grooved margin, and a broad flattened slightly curved beak ½ the length of the rest of the fruit. HABITAT: damp places and ditches; circum-Medit. March–June. A rare British casual.

The similar *R. parviflorus L.* has reflexed sepals, rounded-lobed leaves and warty fruits. A pale greenish yellow plant with softly hairy leaves, which are 3–5-lobed. Flowers smaller, 3–6 mm. across, pale yellow. Carpels small, 2½–3 mm. long, with a narrow border and with shortly hooked swellings over the reddish brown faces. HABITAT: fields, track sides and damp places; Spain to Turkey, N. Africa. March–June. Native in Britain.

R. arvensis L., the CORN CROWFOOT (333) is similar to *R. muricatus* in having distinctive spiny fruits. A taller more erect plant with narrower wedge-shaped, deeply cut leaves. Flowers bright lemon-yellow, 4–12 mm. across, sepals spreading. Carpels covered with conspicuous spines which are longest on the raised margin; beak straight or curved 3–4 mm. long. HABITAT: fields; circum-Medit. April–May. Native, or introduced into Britain.

ADONIS

A. annua L. (A. autumnalis L.) 32   PHEASANT'S EYE

An erect annual with pale finely dissected feathery foliage and bright scarlet or yellow terminal flowers set amongst the leaves. Stem branched, 10–40 cm. high, with leaves 3 times cut into linear segments. Flowers 1–2½ cm. across; sepals hairless, 5; petals 8 or more, often black at the base. Carpels conical, pitted and wrinkled, ending in a short straight beak. HABITAT: fields, stony places, vineyards, cultivated ground; circum-Medit. April–June. Naturalized in Britain.

The similar *A. microcarpa DC. (A. cupaniana Guss., A. dentata Del.)* is distinguished by a tooth on the inner face of the carpel. The flowers are very variable, about 1 cm. across, yellow or reddish violet. HABITAT: cultivated ground, fields, hills; Spain, Italy, Greece to Palestine, N. Africa. March–June.

BERBERIDACEAE—Barberry Family

Usually spiny bushes with clusters of small yellow flowers and undivided leaves but some Mediterranean species are herbaceous plants with under-ground swollen tubers and compound leaves. Flowers yellow with petals and sepals often similar in appearance; stamens the same number as the petals, anthers opening by valves; fruit a fleshy berry or splitting capsule. Many are ornamental.

BONGARDIA: leaves arise only from the rootstock.
LEONTICE: stems leafy.

BONGARDIA

B. chrysogonum *(L.) Griseb.* (*Leontice chrysogonum L.*) 33

An unusual-looking plant with yellow flowers borne on branched naked stems, and compound leaves arising direct from the ground with bluish grey often brown-blotched leaflets. A perennial with a spherical corky tuber, 6 cm. or more across. Leaves with paired obovate wedge-shaped leaflets each having 3–6 teeth at the apex. Flowering stem about 20 cm. high, with small inconspicuous bracts, and terminal bright yellow flowers about 1–1½ cm. across with 6 petals. Fruit narrow egg-shaped, 1 cm. long, generally turning reddish on ripening. HABITAT: fields and disturbed ground; Greek Islands to Palestine. February–March.

The tuber is used as a cure for epilepsy.

LEONTICE

L. leontopetalum *L.*

Rather similar in appearance to the last but with many-flowered leafy stems, and much more dissected leaves. A corky tuber below ground bears an erect branched stem, 20 cm. or more high. Leaves blue-grey, divided into 3 main lobes which are further divided into rounded blunt, egg-shaped leaflets. Flower heads branched, flowers yellow, long-stalked and borne in the axils of leafy bracts. Fruit bladder-like, inflated egg-shaped, 2–3 cm. long. HABITAT: sunny rocky places, cultivated ground; Greece to Palestine, Egypt, Algeria. February–April.

The tuber is used as a cure for epilepsy. Sometimes used as soap.

LAURACEAE—Laurel Family

A mainly tropical family of very aromatic trees and shrubs. Leaves alternate, leathery, evergreen, and dotted with shining oil glands. Flowers small, greenish or yellowish, often in clusters; sepals 6; stamens typically in 4 whorls; ovary with 1 ovule; fruit usually a berry. The family produces many useful products such as cinnamon, camphor, sassafras, avocado pears, and timber such as greenheart wood.

LAURUS

L. nobilis *L.* 31   LAUREL, SWEET BAY

A bushy tree, with typical laurel-like dark green leathery leaves which are very aromatic when crushed, and yellowish flowers in small axillary clusters among the leaves. A tree with black bark and erect, closely spaced branches, up to 10 m. high. Leaves elliptic to lance-shaped, 8–10 cm. by 2–3 cm., with smooth margins. Flowers 4–6, clustered in the axils of the leaves, with 4 petals; either male with 8–12 stamens, or female with single ovary and 2–4 sterile stamens. Fruit a black berry. HABITAT: damp rocks and ravines, thickets and old walls; circum-Medit., but often planted. March–April.

The plant yields the fragrant 'oil of Bay'. The tree was much venerated in classical and biblical times, and it was sacred to Apollo and used as a symbol of victory in crowns and garlands of honour. The word baccalaureate refers to laurel berries and to wreaths worn by scholars and poets when addressing the university or receiving academic honours. Commonly grown as a potherb. Theophrastus describes the wood as 'hot' wood used for making firesticks, and also walking sticks and fire drills as it does not wear away quickly.

PAPAVERACEAE—Poppy Family

Herbaceous plants with milky or watery juice and leaves mostly divided and compound. Flowers usually large and conspicuous with 2–3 sepals which fall as the flowers open; petals 4–6, often crumpled at first; stamens numerous; fruit either a top-shaped capsule opening by pores, or a long pod opening by longitudinal slits. Many cultivated and ornamental plants; opium is an important drug.

PAPAVER: annuals or perennials with milky juice; flowers with 2 sepals which drop early; fruit with a cap of radially arranged stigmas, opening by pores below the cap.

ROEMERIA: poppy-like, but in our species flowers violet; fruit long and narrow of 3–4 valves.

GLAUCIUM: flowers large, solitary, yellow or red; sepals 2; petals 4; fruit of a long curved narrow pod opening into 2 valves.

HYPECOUM: flowers small, yellow, in few-flowered spreading heads; petals 4, unequal in size, stamens 4, and fruit forming a long narrow pod which divides into many 1-seeded sections; leaves much divided into narrow segments.

FUMARIA: flowers small, in clusters, petals 4, the upper hooded and spurred, the lower boat-shaped, and 2 narrow lateral petals; fruit 1-celled and usually 1-seeded. Sometimes placed in the family *Fumariaceae*.

PAPAVER

P. rhoeas L. 34   CORN POPPY

The commonest cornfield poppy, distinguished by its globular flat-topped hairless seed-pod. A hairy annual, 25–90 cm. high, with white juice and once or twice cut leaves with lance-shaped acute segments. Flowers large, 7–10 cm. across, deep scarlet, often with dark blotch at base of the petals; flower stalk with conspicuous spreading bristles. Pod as long as broad, with 7–10 radiating stigmas. HABITAT: harvest fields, waste places; circum-Medit. April–July. Native in Britain.

A very variable plant with the E. Mediterranean forms appearing much darker coloured than the British plants; it is the parent of the Shirley Poppies of gardens. Seeds are used as a tonic for horses by the Arabs and Turks, and infusions from the fruits are good for coughs, also as an eye lotion for animals. A red ink is obtained from the petals

74

The similar *P. hybridum L.*, the BRISTLY POPPY, has more or less globular seed pods covered, with long yellowish brown spreading bristles. Leaves and stems with spreading or flattened hairs. Flowers rather small, 2–5 cm. across, their stalks with flattened hairs; brick-red or wine coloured, with purplish-black patch at the base of each petal. Pod 1–1½ cm. long, at most twice as long as broad, with 4–8 radiating stigmas. HABITAT: cultivated ground, cornfields, waste places; circum-Medit. March–July. Native in Britain.

P. somniferum L.   OPIUM POPPY
A conspicuous annual with grey waxy leaves clasping the stem, and large floppy usually violet flowers. A robust annual, 40–100 cm. high, with broad and coarsely toothed wavy-margined leaves, the upper clasping the stem. Plant usually hairless, but *var. setigerum (DC.) Thell.* has spreading bristly hairs on flower-stems and leaves. Flowers very large, 10–18 cm. across, sometimes white, usually with a dark basal blotch; petals sometimes deeply cut. There are many cultivated forms of Opium Poppy. HABITAT: fields, roadsides and waste places, often escaping from cultivation; circum-Medit. April–July. Casual in Britain.

Young capsules contain many alkaloids, collectively known as opium, which include the narcotics morphine and codeine. Opium latex is collected from incisions made in young green but fully grown seed pods; when smoked or chewed it causes intoxication which in excess may be fatal. Poppy seed oil contains no narcotic properties and is used as a salad oil. This plant was known to the Greeks but is not recorded in the Bible.

ROEMERIA
R. hybrida (L.) DC.
A poppy-like plant with large violet flowers and long cylindrical fruits. A branched hairy annual, 20–40 cm. high, with yellow sap. Leaves 2–3 times divided into linear hair-pointed segments. Flowers 4–5 cm. across, shortly stalked, falling by mid-day. Pod 5–7 cm. long and 2–3 mm. broad, with spreading hairs. HABITAT: fields and hilly places; circum-Medit. (not Italy). March–June. A very rare British casual.

GLAUCIUM
G. flavum Crantz   YELLOW HORNED-POPPY
A conspicuous plant with large irregular bluish grey leaves, large yellow flowers and very long curved fruits. Stem erect, branched, 30–90 cm. high, bearing thick rough bluish-grey leaves; root leaves deeply lobed with toothed segments, upper leaves clasping the stem, deeply lobed or wavy-edged. Flowers 6–9 cm. across, golden yellow or pale yellow. Pod hairless but rough, 15–30 cm. long. HABITAT: shingle and coarse sand on the littoral, waste places, limestone hills; circum-Medit. May–August. Native in Britain.

The similar *G. corniculatum (L.) Rudolph*, the RED HORNED-POPPY (35), has smaller scarlet or orange-red flowers, often with a black spot at the base of each petal. Stem shorter, 25–30 cm., with spreading hairs; leaves similar, rough-hairy. Flowers 3–5 cm. across. Pod 10–22 cm. long, slightly curved, rough-hairy. HABITAT: cultivated fields and waste places; circum-Medit. April–June. A British casual.

HYPECOUM

H. procumbens L. 36
A rather delicate annual with blue-grey dissected leaves, orange juice and small yellow flowers in loose heads. Stems spreading or ascending, 10–40 cm. long, and leaves 2–3 times dissected into ovate-lance-shaped segments. Flowers ½–1½ cm. across, on short stalks with conspicuous bracts; petals unequal, the outer broadly rhomboidal with 3 terminal lobes, the lateral lobes blunt-oblong, and the central lobe triangular. Pod somewhat swollen at the joints. HABITAT: fields, cultivated ground, dry stony places; circum-Medit. March–April.

The similar *H. grandiflorum Benth.* has often somewhat larger, deeper orange-yellow flowers ranging from 1–1½ cm. across, and the outer 2 petals are broadly triangular, wedge-shaped at the base, with 3 equal triangular lobes at the apex. Pod curved, hardly swollen at the joints. HABITAT: hillsides and fields; France to Palestine, Egypt. February–April.

FUMARIA

F. capreolata L. 37   RAMPING FUMITORY
A large-flowered fumitory with clusters of cream-coloured flowers with dark crimson tips, and long climbing stems. An annual with bright green 'ferny' leaves and flexuous stems climbing up to 1 m. through supporting vegetation. Leaves much divided into pointed lance-shaped segments. Flowers conspicuous, 11–15 mm. long, clustered into rather dense flower-heads, about 20-flowered. Fruits on curved stalks, smooth and spherical with a flattish top. A variable plant with several colour forms. HABITAT: hedges, stony places, walls and cultivated ground; circum-Medit. March–June. A British native.

CAPPARIDACEAE—Caper Family

Herbs, trees and shrubs with alternate simple or compound leaves, with stipules, and in our species, spiny. Sepals and petals usually in 4's, stamens 6–many. Ovary in our species on a long stalk and carried above the other floral parts. Fruit a capsule or berry.

CAPPARIS

C. spinosa L. (including *C. rupestris S. & S.*) 384   CAPER
A straggling spiny shrub with large solitary pinkish flowers with numerous long stamens. Stems slender, spreading, up to 1½ m. long;

leaves rather fleshy, oval, with 2 thick reflexed spines at the base of the short leaf-stalk. Flowers 4–6 cm. across on thick stalks; stamens much longer than the 4 rounded petals. Ovary long-stalked; fruit a berry 5 cm. long. HABITAT: walls, rocks; circum-Medit. May–July.

The flower buds are pickled to make capers.

Var. inermis Turra. has pendulous branches; leaves fleshy, hairless and dark green, more or less rounded. Stipules hardly spiny, straight and soon dropping off so that the plant appears spineless. HABITAT: walls, rocks; probably circum-Medit. May–July.

CRUCIFERAE—Mustard Family

Herbaceous plants with alternate leaves without stipules, and branching heads of many small flowers usually without bracts. Flowers very uniform in structure; sepals 4, petals 4, arranged in a cross – hence the name 'crucifer'; stamens usually 6 and fruit of 2 fused carpels. Fruit usually a pod, either long and narrow, a *siliqua*, or if not more than twice as long as broad, a *silicula*. Fruit usually splitting longitudinally into 2 valves, but sometimes splitting transversely into several 1-seeded sections. A large and important family which includes the cabbages, cauliflower, kale, kohlrabi, turnip, radish, watercress, as well as mustard and horse-radish. Many are weeds of cultivation and many are grown ornamentally.

BUNIAS: flowers white or yellow; fruit ovoid with warty or wing-like crests, not splitting.

CHEIRANTHUS: flowers yellow or orange, sometimes tinted with brownish purple; pod elongated, somewhat flattened and more or less 4-sided; leaves simple and undivided.

MALCOLMIA: flowers violet, purple or white; fruit a long narrow cylindrical pod with 1 row of seeds, and with each valve with 3 veins; leaves undivided.

MATTHIOLA: plant greyish and covered with branched hairs, and flowers purple, red or white; fruit a long cylindrical pod ending in terminal horn-like processes, each valve 1-nerved.

ALYSSUM: low-growing mat-forming herbs with simple leaves and stems usually covered with star-shaped hairs; flowers yellow or white; fruit disk-shaped, faintly netted on the flat sides.

LOBULARIA: similar to *Alyssum* but flowers white and whole plant covered with forked, not star-shaped hairs.

DIPLOTAXIS: flowers yellow, white or lilac; leaves deeply divided; fruit a long slender short-beaked pod with flattened 1-veined valves and seeds in 2 rows.

SINAPIS: annuals with yellow flowers and spreading sepals; fruit a long pod of 2 round-backed valves each with 3–7 veins and with a long beak.

ERUCA: flowers usually yellowish with violet veins; fruit a short siliqua with valves strongly 1-veined and a broad flat seedless beak.

MORICANDIA: flowers large, violet and veined; fruits long, narrow and 4-angled with a short beak; valves 1-nerved and seeds winged.

CARDARIA: flowers small, white, in dense flattened clusters; fruit broadly egg-shaped to heart-shaped, 2-seeded and not splitting when ripe.

BISCUTELLA: flowers small, yellow; pods of 2 circular disk-like lobes placed edge to edge which are flattened and 1-seeded.

AETHIONEMA: flowers white or violet in showy clusters; usually with glaucous undivided hairless leaves; fruit flattened, winged, rounded in outline, usually with a notch at the apex, each valve usually many-seeded.

BUNIAS

B. erucago L. 334 (Fig. IIb)
A crucifer with small yellow flowers and distinctive warty, winged and toothed pods. A branched annual or biennial, 30–60 cm. high, with rough glandular-hairy stems and leaves. Lower leaves deeply cut into broad blunt lobes, the upper stalkless, oblong, toothed or entire-margined. Flowers 1 cm. or more across, petals heart-shaped, sepals spreading. Pods about 1 cm. long, 4-angled with 4 broad-toothed and jagged interrupted flanges, and a style ½ cm. long. HABITAT: fields and roadsides; Spain to Turkey, Algeria. May–July. A British casual.
The roots and shoots are eaten as salad in Greece.

CHEIRANTHUS

C. cheiri L. (Fig. IId)   WALLFLOWER
This well-known plant with its showy yellow or orange-brown very sweet-scented flowers and narrow leaves is now widely naturalized in the Mediterranean region. A short-lived perennial, somewhat woody at the base with erect leafy stems, 20–80 cm. high. Leaves stiff, untoothed, narrow lance-shaped, covered with forked hairs pressed to the blade. Flowers large, 2 cm. across, in dense showy spikes. Pods erect, rectangular, flattened in section, 3 mm. broad, each valve 1-veined. HABITAT: walls, ruins and rocks, often escaped from gardens; probably native of E. Medit. March–May. Introduced and well established in Britain.

MALCOLMIA

M. maritima (L.) R.Br. 38, 335 (Fig. IIa)   VIRGINIAN STOCK
A low-growing annual with bright rosy-purple or lilac flowers and long thin spreading fruits. Sparingly branched from the base, 10–40 cm. high. Leaves few, elliptic to lance-shaped, toothed. Flowers 1½ cm. across, white- or orange-centred, petals notched. Fruits stalked, up to 6 cm. long, about 2 mm. broad, and covered with flattened hairs (some branched into 3) and

ending in a long pointed hairless style. HABITAT: dry places, particularly near the sea, waste places near habitation; Spain to Greece, naturalized in N. Africa. April–June. Grown as an ornamental plant.

The similar *M. crenulata* (*DC.*) *Boiss.* has much longer pods, 8–12 cm. long, and leaves which are narrow lance-shaped with shallow rounded teeth. Flowers 1½ cm. across, pink (white in the Judaean hills). Pods more or less hairless. HABITAT: common in deserted fields, vineyards; Turkey to Palestine. January–May.

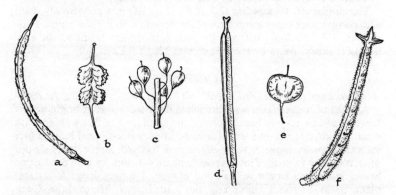

Fig. II. Fruits of *Cruciferae*
a *Malcolmia maritima* (×1) b *Bunias erucago* (×1½) c *Lobularia maritima* (×1½) d *Cheiranthus cheiri* (×1) e *Alyssum saxatile* (×1½) f *Matthiola tricuspidata* (×1)

M. flexuosa S. & S. has larger flowers with the fruit-stalks very much swollen and as thick as the pods, and style conical and hardly distinct. Pods 5–8 cm. long, arched, with flattened hairs, stigma 3–4 mm. long; leaves stalked, untoothed. HABITAT: sandy places by the sea; Greece, Turkey, Cyprus. March–June.

MATTHIOLA—Stock
M. sinuata (*L.*) *R.Br.* 42   SEA STOCK
A plant with characteristically grey thick wavy-margined leaves and a loose head of pale rose-lilac flowers. A perennial or biennial, 20–60 cm. high, with spreading branches, very leafy at the base, and covered with glandular hairs. Lower leaves toothed, wavy-margined or deeply indented, upper narrow lance-shaped, entire. Flowers stalked, 2–2½ cm. across, sweet-scented, especially in the evening. Pods 7–12 cm. long, glandular-hairy, with projecting tooth-like lobes spreading in a star at the apex. HABITAT: maritime rocks; European Medit., Asia Minor and Algeria. March–June. Native in Britain.

M. longipetala (*Vent.*) *DC.subsp. bicornis* S. & S. **336**

NIGHT-SCENTED STOCK

A dull-coloured stock with the upper leaves without teeth or wavy margins. Stem branched from the base, leafy below, up to 30 cm. high. Lower leaves oblong lance-shaped and deeply lobed, upper linear and un-toothed. Flowers stalkless, dull or dark purple, expanding at evening and closed by day, very sweet scented when open. Pods spreading or reflexed, 6–8 cm. long, and ending in 2 spreading horns up to 1 cm. long. HABITAT: cornfields, vineyards; Greece to Palestine. April–May.

The similar *M. tricuspidata* (*L.*) *R.Br.* (Fig. IIf) is a grey-woolly plant with conspicuous sweet scented rose-lilac flowers with a white centre. Pods shorter, 3–6 cm. long, stiff woolly with 3 triangular horns at the end. HABITAT: sandy places by the sea; circum-Medit. March–July.

ALYSSUM

A. saxatile ssp. orientale (*Ard.*) *Beck*   **40**   (Fig. IIe)   GOLDEN ALYSSUM

A plant of rocky places with branched flattened heads of bright yellow flowers, and silver-grey basal leaves. A perennial, up to 30 cm. high, with erect branched stems and leaves covered in grey star-shaped hairs; lower leaves narrowly obovate, wavy-edged and toothed, upper lance-shaped often minutely toothed. Flowers many, small, 4–6 mm. long, in flat-topped heads, golden or lemon yellow. Pod elliptic, 4–6 mm. long. A variable plant. HABITAT: rocks, cliffs; Yugoslavia to Turkey. March–June. Commonly grown in rock-gardens in Britain and sometimes escapes.

LOBULARIA

L. maritima (*L.*) *Desv.* (*Alyssum maritimum* (*L.*) *Lam.*)   **41**   (Fig. IIc)

SWEET ALISON

A small silvery-white-leaved plant with close clusters of tiny white, sweet-scented flowers with the petals noticeably rounded in outline. A perennial with a somewhat woody base but with many spreading her-baceous green stems, 10–30 cm. long. Leaves narrow lance-shaped, silvery-haired and densely scattered along the stem. Flowers about 5 mm. across in flattened heads which elongate as they mature, white or slightly rose-coloured. Pods elliptic, hairy, 2½ mm. long excluding style. HABITAT: sandy and rocky places on the littoral; circum-Medit. April–August. Commonly grown in Britain as an edging plant, and often escaping and becoming locally naturalized.

DIPLOTAXIS

D. erucoides (*L.*) *DC.*   (Fig. IIIb)   WHITE WALL ROCKET

A noxious weed of cultivated ground with white flowers veined with violet. A hairy or nearly hairless annual, 30–50 cm. high, branched from the base with basal leaves deeply cut into broad lobes with coarse irregular whitish horny-tipped teeth, upper leaves stalkless and undivided. Flowers about 15 mm. across; petals twice as long as the sepals. Pods 2½–3 cm. by 2 mm., with a conical beak 2–4 mm. long, borne on stalks shorter than the pods. HABITAT: cultivated ground, fields, vineyards; circum-Medit. January–June. Naturalized in Britain.

SINAPIS

S. arvensis L. (Fig. IIIa)   CHARLOCK

A common and ubiquitous weed of cultivation with bright yellow flowers, rough-hairy, deeply lobed basal leaves, and upper stalkless lance-shaped leaves. An annual with simple or branched stem, 30–80 cm. high; leaves large, up to 20 cm. long, the lowest with a very large toothed terminal lobe. Petals 9–12 mm. long. Pod 2½–4 cm. long with a conical straight beak rather more than ½ as long as the lower part. HABITAT: cultivated ground, fields and waste places; circum-Medit. March–May. Probably native in Britain.

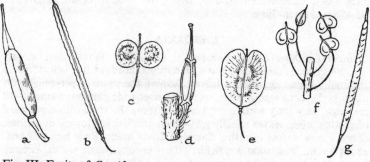

Fig. III. Fruits of *Cruciferae*
a *Sinapis arvensis* (×1) b *Diplotaxis erucoides* (×1) c *Biscutella laevigata* (×1) d *Eruca sativa* (×1) e *Aethionema saxatile* (×1½) f *Cardaria draba* (×1½) g *Moricandia arvensis* (×1)

The similar *S. alba L.*, the WHITE MUSTARD, is distinguished by having all its leaves stalked and deeply lobed, and by the pods with a long flattened sabre-like, often curved beak, as long or longer than the lower part. Petals yellow. HABITAT: fields and rough ground; circum-Medit. March–April. Naturalized in Britain.

Cultivated for fodder and as a green manure. Its seeds give white mustard; commercial mustard is a mixture of seeds of this species and S. nigra L.

ERUCA

E. sativa Mill.   338 (Fig. IIId)

A mustard-like plant with pale yellow or whitish petals, veined with deep violet. A hairy fetid annual, branched from the base, 20–60 cm. high. Leaves deeply lobed with a large broad terminal lobe and 2–5 pairs of broad lateral lobes. Flowers quite large, up to 2 cm. across, with delicately veined egg-shaped petals with long narrow bases. Pods 1–2½ cm. long, by 3–5 mm. broad, erect and pressed against stem with a sabre-shaped beak half as long as the pod. A very variable species. HABITAT: waste ground, fields, olive groves, stony hills and track sides; circum-Medit. February–June. Casual in Britain.

Cultivated as a salad plant; its seeds give a medicinal oil.

MORICANDIA

M. arvensis (L.) DC. 39 (Fig. IIIg)
A distinctive plant with conspicuous violet flowers and smooth rather fleshy rounded grey-green leaves clasping the stem. A hairless biennial with an erect branched stem, 30–40 cm. high, with lower leaves stalked obovate, often shallowly toothed, upper leaves encircling the stem with projecting lobes. Flowers 2 cm. across. Pods narrow, cylindrical. HABITAT: track sides, uncultivated ground, fields; W. Medit., Spain to Greece, N. Africa. March–June.

CARDARIA

C. draba (L.) Desv. (Lepidium draba L.) (Fig. IIIf) HOARY PEPPERWORT
A conspicuous cress-like plant with terminal heads of tiny white flowers and many broad egg-shaped leaves clasping the stem. A perennial with many herbaceous stems, often spreading over wide patches; stems 30–70 cm. high, very leafy with upper leaves clasping the stem with downward projecting lobes, leaves usually grey-green and hairy. Flowers numerous, about 4 mm. across in rather flattened dense heads. Pods heart-shaped, about 4 mm. long. HABITAT: fields, fallow ground, waste places; circum-Medit. March–June. Naturalized in Britain. A troublesome weed spreading by root-buds and seeds.

BISCUTELLA

B. laevigata L. 337 (Fig. IIIc) BUCKLER MUSTARD
A mustard-like plant with yellow flowers distinguished by the twin, flattened disk-shaped valves of the ripe fruit resembling a pair of spectacles. A hairy perennial, 20–60 cm. high, with root leaves in a rosette and few small stalkless stem-leaves. Leaves very variable, oblong, deeply or shallowly toothed, with or without hairs. Flowers pale yellow, petals 6 mm. long, twice the length of the sepals. Fruit with winged membraneous margins, usually hairless and with a short style. A very variable plant. HABITAT: rocks, dry banks, stony places, open woods; W. Medit., Spain to Yugo-slavia, Morocco. March–July.

B. didyma L. is an annual with similar fruits which have a conspicuous margin but not winged; fruits with short hairs on the margin and rough faces. HABITAT: stony places, dry hills; Greece to Palestine, Egypt. March–May.

AETHIONEMA

A. saxatile (*L.*) *R.Br.* (Fig. IIIe)   BURNT CANDYTUFT

A small spreading plant with bluish-grey leathery leaves, rounded heads of pale purplish flowers and flattened oval pods which are broadly winged and notched at the apex. A short-lived perennial with a woody base and leafy stems, 10–30 cm. long. Leaves thick, hairless, oval to lance-shaped. Flowers pale flesh-coloured or lilac. Pods many-seeded, heart-shaped, 5–7 mm. long, broadly winged towards the apex; style very short, free part $\frac{3}{10}$ mm. long and much shorter than the notch; some pods 1-seeded. HABITAT: stony and rocky ground, mainly on limestone in the foothills; Spain to Greece, N. Africa. April–June.

Var. graecum Boiss. & Heldr. (**43**) has oblong lance-shaped sharp-pointed leaves, and an obovate fruit which is notched, with a style as long as the notch and the free part $\frac{3}{10}$–1 mm. long. HABITAT: rocks and cliffs; Yugoslavia and Greece. April.

RESEDACEAE—Mignonette Family

Annual or perennial herbs with alternate simple or divided leaves, with glands in the place of stipules. Flowers in a close spike; flowers asymmetrical; sepals 4–7; petals 4–7 with those at the back larger and more deeply divided into narrow lobes than those in the front; stamens 7–40, borne on a fleshy disk and crowded to the front. Ovary of 2–6 partially fused or free carpels; fruit a splitting capsule.

RESEDA

R. lutea L. **44**   WILD MIGNONETTE

A pale green cut-leaved plant with long narrow spikes of greenish-yellow flowers. An annual or perennial, 20–60 cm. high, with a rough, ribbed stem. Leaves deeply lobed into 1–3 pairs of narrowly oblong blunt lobes which may be further lobed; lobes with curly crisped margins. Flowers 6 mm. across in compact conical spikes; petals and sepals usually 6. Ripe pods erect, 1–1½ cm. long with 3–4 teeth. HABITAT: fallow ground and stony places, banks and tracks, especially on limestone; circum-Medit. April–July. Native in Britain.

The similar *R. alba L.*, WHITE or UPRIGHT MIGNONETTE (**45**), has spikes of white flowers. Leaves deeply lobed with 5–8 pairs of narrow unequal lobes with more or less undulating margins. Flowers larger, 9 mm. across, with usually 5 sepals and petals, petals 3-lobed. Pods 12 mm. long, erect, with 4 teeth. HABITAT: dunes, sand, fields, roadsides and walls; circum-Medit. March–June. British casual.

CRASSULACEAE—Stonecrop Family

Usually herbs with fleshy leaves, without stipules. Parts of flowers in whorls, varying from 3's to 32's, but usually in 5's. Carpels the same number as the petals, superior and free or united at the base, many-seeded.

SEDUM—Stonecrop

S. stellatum L. 365

An annual stonecrop with broad, toothed fleshy leaves and star-shaped pink flowers. A small plant 5–15 cm. high, with stout ascending stems. Leaves 1–1½ cm. long, flat, toothed, egg-shaped and prolonged into a short leaf stalk. Flowers in axils of leafy bracts ranged along one side of a forked stem. Petals pointed, lance-shaped, hardly longer than the narrow blunt sepals; fruit opening into a spreading star. HABITAT: rocks and stony places; Spain to the Isles of Greece and Crete. May–June.

S. sediforme *(Jacq.) Pau (S. nicaeense Allioni, S. altissimum Poir.)*

A pale yellow-flowered stonecrop bearing a flat-topped flower head of outwardly curving branches. A hairless, bluish-grey perennial with flowering stems 30–50 cm. high and with non-flowering branches of densely overlapping leaves in 5 ranks. Leaves large, swollen, straight, oval to lance-shaped and pointed. Flowers with 5–8 spreading blunt petals, 2–3 times longer than the oval blunt sepals. HABITAT: rocks, walls and stony places, largely on calcareous soils and clay; European and Asian Medit. June–August.

PLATANACEAE—Plane-tree Family

A family of a single genus of trees with lobed leaves and sheathing stipules which fall off when the buds burst. The flowers are 1-sexed in spherical heads with insignificant petals and sepals; male flowers of many stamens; female flowers in separate heads of small pyramidal ovaries. Fruit a rounded head of 1-seeded ovaries surrounded at the base by long hairs. The bark of the trunk flakes off in patches.

PLATANUS

P. orientalis L. 46   ORIENTAL PLANE

A large tree with straight trunk and massive branches and mottled yellowish-green bark, which is commonly planted as a shade tree in many villages and towns in the Mediterranean. Leaves 18–25 cm., about as broad as long and cut to beyond the middle into 5–7 deep lobes (with narrow sinuses) with the middle lobe much longer than broad; mature leaves hairless. Flowers in a 1-sexed catkin composed of globular clusters which are widely spaced on pendulous stalks. HABITAT: in forests in the hills, by rivers; probably native in Turkey but widely planted and sub-spontaneous elsewhere in the Mediterranean. March–May.

The similar LONDON PLANE, *P.* × *hybrida Brot. (P.* × *acerifolia (Ait.) Willd.)* is a hybrid which arose spontaneously about 1700. It is now more commonly planted in Europe than the Oriental Plane. It is distinguished by the leaves which are cut to less than the middle of the blade, by the wide shallow sinuses between the lobes, and by the middle lobe which is nearly as broad as long. The Oriental Plane is one of the parents, the American species, *P. occidentalis L.*, the other.

The Oriental Plane is the famous 'Chenar' of the Persians and Moghuls and was planted for ornament and shade in their gardens. The Greeks and Romans prized the Plane tree for its shade and beauty. Pliny was surprised that a tree should be brought from so far for its shade alone, and Xerxes was so impressed with the plane trees in Lydia that he decorated them with golden ornaments. The wood takes a fine polish and is much prized by cabinet makers.

ROSACEAE—Rose Family

A family of trees, shrubs and herbaceous plants of the temperate regions. Leaves simple or compound, with stipules. Flowers with petals and sepals usually 5; stamens usually 10, 15 or 20. Ovary variable, superior, half-inferior, or inferior; fruit of many forms, fleshy or dry – a berry, pome, drupe or achene. Many important fruits come from this family such as apple, pear, plum, cherry, peach, apricot, almond, strawberry, raspberry, blackberry and loquat.

POTERIUM: leaves compound; flowers 1-sexed in dense spikes; sepals 4, green, stamens numerous, stigmas feathery.

ERIOBOTRYA: leaves large, strongly veined; flowers hermaphrodite in terminal branched clusters; fruit a pome with 1–2 large seeds.

PYRACANTHA: thorny shrubs with serrated leaves and conspicuous stipules; flowers in flat heads; fruit with 5 stones.

CRATAEGUS: thorny trees or shrubs with lobed leaves and small stipules; flowers in flat-topped heads; carpels 1–5, free at the apex but fused at the base; fruit with 1–5 stones.

PRUNUS: flowers solitary or in clusters; stamens usually 20; ovary with 1 carpel in a cup-shaped receptacle; fruit a 1-seeded drupe.

POTERIUM

P. spinosum L. **48**   THORNY BURNET

A compact, almost hemispherical, very spiny and intertwined shrub which is a dominant plant of large areas of garigue in the Eastern Mediterranean. Usually about ½ m. high; leaves small with 8–14 leaflets which fall in the summer; the young branches end in double thorns which gradually harden and turn brown. Flowers 1-sexed, in compact oval heads; male heads with long yellow stamens; female heads with purple feathery styles. Fruits swollen and spongy, and sometimes turn a bright red as they develop. HABITAT: dry limestone hills, particularly where forest and maquis have been cleared; abandoned cultivation; Eastern Medit. from Italy, Sicily and Sardinia to Palestine. March–April.

This shrub is commonly cut in summer for fuel particularly for use in bake-ovens and lime kilns; in Arab villages the drying branches often line the tops of enclosure walls. Probably not the plant used for Christ's 'crown of thorns'.

85

ERIOBOTRYA

E. japonica (Thunb.) Lindl. 47   LOQUAT

A native of Japan very commonly cultivated in the Mediterranean region. It is a small ornamental tree, 3–10 m. high, with large oval, horny, strongly veined leaves of a clear dark green. Branches and underside of leaves covered in rust-coloured hairs. Its white flowers open in winter and are sweet-scented. The fruits are about the size of a plum, golden yellow, and ripen from the end of April to June; they are very juicy, sweet and somewhat acid, but contain large seeds. Introduced to the Mediterranean region at the beginning of the nineteenth century; it is easily cultivated and requires little attention.

PYRACANTHA—Firethorn

P. coccinea Roemer

A dense spiny evergreen shrub, 1–2 m. high, with flat heads of white flowers and scarlet fruits. Leaves elliptic, 2–3 cm. by 1–1½ cm., with rounded teeth, hairless and shining above, hairy below when young. Flowers white or pinkish-yellow, clustered into many-flowered conspicuous umbels; petals scarcely longer than the hairy sepals. Fruit the size of a pea, scarlet, and persisting throughout the winter. HABITAT: woods and hedges; Spain to Turkey. May–June.

Often grown as a hedge plant and commonly escaping from cultivation.

CRATAEGUS

C. azarolus L.   MEDITERRANEAN MEDLAR

Probably the most widespread hawthorn in the Mediterranean region; distinguished by the whitish cottony hairs on the young branches, leaf stalks and flower stalks. A spiny shrub or small tree, 4–12 m. high, with egg-shaped leaves, wedge-shaped at the base and deeply divided into 3–5 oblong-toothed lobes, hairy on both surfaces. Flowers white in flat heads; styles 1–3. Fruit red or yellow, large, 1–1½ cm. in diameter. The fruits are edible and have a pleasant but acid taste. HABITAT: woods, hedges and field verges; France to Palestine, Algeria. April–May.

The similar HAWTHORN *C. monogyna Jacq.* has hairless young shoots; smaller deep red fruits, 8–10 mm.; style 1. HABITAT: bushy places; circum-Medit. May.

PRUNUS

P. communis (L.) Fritsch (P. amygdalus Batsch)   ALMOND

The almond is widely cultivated in the Mediterranean region and often escapes and becomes naturalized. A tree, 4–10 m. high, with green branchlets. Leaves lance-shaped, bluntly toothed, with glands on the short leaf-stalk. Flowers pink, sometimes almost white, 3–5 cm. across, appearing before the leaves. Fruit green, dry and leathery in texture; nut deeply wrinkled. HABITAT: cultivated ground, thickets, hedges and rocky places near cultivation; circum-Medit. February–March.

There are several varieties. Var. *dulcis produces a sweet edible oil, and* var. *amara produces a bitter oil which is narcotic and contains prussic acid. The semi-wild trees are usually bitter. The wood is reddish in colour and used for veneering.*

LEGUMINOSAE—Pea Family

A very large and important family of herbs, trees and shrubs. Leaves usually compound, often trifoliate and rarely simple, usually with stipules. Many are climbing plants and show leaves or parts of leaves modified to tendrils. Flowers with petals and sepals commonly in 5's, stamens 10 or sometimes numerous, ovary 1; many are very characteristically 'butterfly-like'. Fruit a legume – a 1-celled, many-seeded pod which splits into 2 separate valves.

The *Leguminosae* is divided into 3 sub-families. The *Mimosoideae* have many small clustered flowers with numerous conspicuous stamens protruding beyond the small petals; the flowers are symmetrical. The *Caesalpinioideae* usually have large butterfly-like flowers and unfused stamens. The *Papilionatae* also have butterfly-like flowers with an upper petal or *standard*, 2 lateral petals or *wings* and a *keel* formed of 2 lower petals; sepals 5, fused into a tube; stamens 10, usually enclosed in the petals, fused by their stalks into a tube which encloses the ovary, or with 1 stamen free.

A family of great importance to man for it supplies food, fodder, dyes, gums, resins, oils and many timbers. It has nodules on its roots which contain bacteria able to fix nitrogen, and in consequence a member of this family is regularly included in crop rotation. Of particular importance are: lentils, peas, beans, ground-nuts, soy beans, indigo, alfalfa and clover. There are many ornamentals.

ACACIA: trees; flowers usually numerous in clusters; leaves compound or reduced to a flattened leaf-like stalk.

CERATONIA: a tree; flowers green, petals absent, stamens 5; leaves of 4–8 large leathery evergreen leaflets; pods large.

CERCIS: trees; petals reddish-purple; stamens 10; leaves undivided, rounded with heart-shaped base.

ANAGYRIS: a fetid bush; leaves trifoliate; flowers yellow; stamens not fused by their stalks.

LUPINUS: leaves with leaflets arranged like the fingers of a hand (palmate); flowers conspicuous in long terminal spikes.

COLUTEA: shrubs without spines; leaflets paired with a terminal leaflet; pod membraneous, much inflated and bladder-like when ripe.

CYTISUS: non-spiny shrubs; calyx divided into 2 diverging lips; leaves usually trifoliate.

CALYCOTOME: shrubs with terminal branches transformed into spines; leaves trifoliate; calyx with short, more or less equal, teeth.

GENISTA: shrubs usually with spines; leaves usually simple or reduced to a spine; calyx 2-lipped, upper lip deeply divided.

RETAMA: shrubs very similar to *Genista* but with white flowers; stems more or less leafless, without spines; fruit swollen, not splitting.

SPARTIUM: spineless shrub with smooth rush-like branches; leaves simple or absent; calyx with a single oblique lip.

ONONIS: leaves simple or trifoliate with veins ending in teeth and with conspicuous stipules fused to leaf-stalk; calyx bell-shaped with 5 more or less equal teeth; flowers rose or yellow; sometimes shrubby.

MEDICAGO: pod usually spirally coiled or less commonly sickle-shaped; flowers orange, petals falling early; leaves trifoliate with stipules; rarely shrubby.

TRIGONELLA: pod straight or slightly curved, narrow or egg-shaped, smooth, often flattened; flowers numerous in short clusters; leaves trifoliate with stipules.

MELILOTUS: flowers small and many in narrow elongated spikes; leaflets toothed at the tip; pod straight, thick and shorter than calyx.

TRIFOLIUM: flowers in dense rounded or hemispherical heads; pod very small, enclosed in calyx; flowers pink, white or purple, not blue; leaflets often toothed.

ANTHYLLIS: calyx long-tubed and inflated at the base; pod enclosed in calyx, 1–3-seeded; flowers in dense flattened heads usually surrounded by conspicuous leafy bracts.

PHYSANTHYLLIS: similar to *Anthyllis* but calyx much inflated and bladder-like; upper stamen not fused to the remaining 9 fused stamens.

HYMENOCARPUS: pod rounded, flattened, disk-shaped with a spiny-toothed margin; leaves of 2–3 pairs of leaflets with terminal leaflet much larger; flowers orange.

BISERRULA: pod like a 2-edged saw, flattened with deeply indented margins; flowers whitish, blue tipped; leaves of 7–15 pairs.

LOTUS: leaves with 5 leaflets; pod narrow elongated, divided between the seeds into sections and splitting into 2 valves; flowers yellow or orange.

TETRAGONOLOBUS: similar to *Lotus* but with 4-angled, winged pods; leaves trefoil with large leafy stipules.

BONJEANIA: similar to *Lotus* but with a straight and blunt keel; leaflets oblong or wedge-shaped; flowers whitish.

ASTRAGALUS: pod 2-valved and often divided longitudinally into 2 cells; flowers conspicuous, usually white, pink or bluish; keel blunt. A large genus of which only an example is given.

SCORPIURUS: pod spiral, 8-sided, covered with spines or swellings; leaves undivided, large and stalked; flowers yellow.

PSORALEA: flowers bluish, in short dense rounded spikes; leaves trefoil with lance-shaped, untoothed leaflets; pod small, ovoid, 1-seeded.

ROBINIA: trees with spines; flowers in hanging white clusters; pods long, flattened and straight-sided.

CORONILLA: pods narrow linear, cylindrical in section, with a short beak and dividing into 2 or more sections when ripe; flowers yellow, purple or white. Sometimes shrubs.

HIPPOCREPIS: pod flattened, narrow and deeply indented into horse-shoe-shaped segments; 2 upper teeth of calyx fused; flowers yellow.

HEDYSARUM: pod broad, flattened and jointed into rounded 1-seeded segments which are veined or covered with spiny hairs; flowers in close clusters, rose or violet, sometimes whitish.

ONOBRYCHIS: pod hard, not splitting or jointed, often with swellings or spiny; flowers usually pink.

VICIA: pod flattened, parallel-sided and splitting into 2 valves; leaves of many pairs of leaflets, usually with a terminal simple or branched tendril; style either hairless or equally hairy all round. A large genus.

LATHYRUS: similar to *Vicia* but usually with fewer leaflets and with winged or angled stems; style with hairs on the upper side only; flower heads usually long-stalked.

PISUM: similar to *Lathyrus* but stems rounded, not winged; stipules larger than leaflets; wings of flower fused to the keel.

ACACIA—Wattle, Mimosa

Many species of Mimosa are cultivated in gardens and by roadsides throughout the Mediterranean, and they often become semi-naturalized in waste places near habitation, for their seeds germinate easily. Mimosas grow well on light sandy soil in warm airy places, and are able to stand the long drought of the Mediterranean summer; they do not usually grow well on calcareous soils. All young seedlings have compound, twice-cut (bipinnate) leaves, but as the plant matures the leaf-stalks of many species develop and flatten, and take over the work of feeding, and the compound leaflets disappear; these flattened blades are technically called phyllodes. Most of the cultivated species are of Australian origin.

A. armata R. Br.   Kangaroo Thorn

A spreading spiny shrub, 3–4 m. high. 'Leaves' simple, ending in a sharp spine, blade vertical, 2½ cm. long, ½-egg-shaped with the straight edge

89

against the stem and the outer edge wavy; stipules composed of slender spines, 6 mm. long. Flower heads solitary, shortly stalked, globular, yellow.

A. cultriformis Hook.
A tall shrub with broad grey-green 'leaves' closely arranged along the branches. 'Leaves' nearly as broad as long, triangular, pointed. Flower heads small, spherical on branched flower stalks much longer than the leaves and bunched together at the ends of the branches.

A. longifolia (Andr.) Willd. 339   WHITE SALLOW, SYDNEY GOLDEN WATTLE
A tall shrub or small tree with narrow, lance-shaped, willow-like 'leaves', 5–15 cm. long and about 1 cm. wide, narrowed at the base. Flowers bright yellow in cylindrical stalkless catkin-like clusters, 2–5½ cm. long, in the axils of 'leaves'; parts of flowers in 4's.
A rapidly growing species which will grow on sand and help to stabilize it. Bark used for tanning; the tree also produces a light tough hard timber used for tool handles, etc.

A. retinoides Schlecht.
A handsome tree with crowded narrow, bright green willow-like 'leaves' which are variable in colour and shape. Flower clusters small, globular, very pale yellow, sweet scented, in bunched clusters at the ends of flexible branches. A rapidly growing tree which, unlike most other species, does well on calcareous soils. Widely planted, it flowers most of the year; one of the first species to be introduced to the Riviera.

A. cyanophylla Benth. 50   BLUE-LEAVED WATTLE
A shrub or small tree up to 6 m. high, with long narrow pointed 'leaves' which are bluish-green in colour, 15–30 cm. long, rather abruptly narrowed at the base. Flower heads golden-yellow, about 1 cm. across, on short stalks, in axillary clusters of 3–5 which are shorter than the 'leaves'.

A. melanoxylon R. Br.   BLACKWOOD ACACIA
A dense pyramidal tree, 8–10 m. high. 'Leaves' sometimes compound but usually simple, dark green, lance-shaped, 6–12 cm. long and up to 2 cm. wide, with one margin more strongly curved than the other. Flower heads about 8 mm. across, cream-coloured, unscented, in axillary clusters shorter than the 'leaves'. It gives a heavy shade and does well on both siliceous and calcareous soils.

A. pubescens R. Br.   HAIRY WATTLE
A shrub, 2–3 m. high, with drooping velvety white branches. True leaves also velvety white, compound, twice pinnate, of 3–10 pairs of segments each bearing 6–20 pairs of narrow leaflets. Flower heads yellow, on bunched, long-stalked clusters, longer than the 'leaves'. One of the most handsome species of wattle.

A. dealbata Link 340   SILVER WATTLE
The most handsome of all the wattles cultivated in the Mediterranean

region, with smooth whitish trunk and long, pendulous, leafy, whitish branches. A tree, 8–12 m. high, with silvery-grey compound leaves, with 13–25 pairs of segments each bearing 30–40 pairs of leaflets. Flower heads deep yellow, with delicate scent, borne in much-branched clusters, largely at the ends of the branches. A rapidly growing species which regenerates naturally by suckers.

A. decurrens Willd.   GREEN WATTLE
Very similar to the Silver Wattle but it is a smaller tree 5–8 m. high, and has reddish, hairless, ribbed branches and dark green compound leaves of 8–15 pairs of segments with 30–40 or more pairs of leaflets each 2–3 mm. long. Flower heads 20–30 flowered in long, little-branched axillary clusters.

A. farnesiana Willd.   POPINAC, OPOPANAX, CASSIE, HUISACHE
A much-branched shrub, 2–3 m. high, armed with straight spines. True leaves of 5–8 segments each bearing 10–25 pairs of narrow leaflets, 2–4 mm. long. Flower heads large, deep yellow and very fragrant, in groups of 2–3 towards the ends of the branches.
Cassie is a perfumed oil obtained from the flowers of this species, used in making scents with 'violet' fragrance.

CERATONIA—Carob

C. siliqua L.   341   CAROB TREE, LOCUST TREE
One of the most characteristic trees of the dryer parts of the Eastern Mediterranean region. Its stout trunks, spreading branches, and dark green leathery foliage give welcome shade throughout the year. The long broad pods, known as locust beans, hang from the branches conspicuously in spring and summer. A large spreading tree, 7–10 m. high, with compound leaves of rounded leathery shining leaflets, dark green above and paler below. Flowers tiny, greenish, grouped in shortly stalked spikes among the leaves, either all male (reddish) or hermaphrodite. Pods very large, 12–20 cm. by 2 cm., thick, leathery and pulpy, turning brown on ripening. HABITAT: commonly cultivated, particularly in Cyprus and E. Medit., but occurring naturally in rocky places on the littoral; circum-Medit. August–October; pods often occur with flowers.
The sugary unripe pods are eaten by the poor and used to feed cattle; they are the 'husks' of the parable of the prodigal son, and probably the 'locusts' eaten by John the Baptist. They are also used for making sherbets and fermented drinks. The seeds were the original 'carat' weight of the jewellers. The wood is hard and lustrous and used for marquetry and walking sticks.

CERCIS—Judas Tree

C. siliquastrum L.   54   JUDAS TREE
A familiar tree of the Mediterranean region with clusters of beautiful purplish rose flowers borne in the early spring often on bare branches before any leaves appear. A small deciduous tree, 5–9 m. high, with finely fissured

bark. Leaves undivided, rather broadly kidney-shaped or orbicular, smooth and light green. Flowers borne on old wood and even from the trunk itself; flowers 2 cm. long; calyx cup-like with blunt teeth, reddish-purple; stamens 10, not fused. Pods hanging, 7–10 cm. by 1½ cm., reddish brown. HABITAT: much cultivated in the Medit. region but native in rocky hills in the E. Medit.; Greece to Palestine. March–April.

This tree is traditionally the tree that Judas hanged himself on; it was said that the pale flowers blushed rose-red for shame. The wood is beautifully veined, very hard, and takes a high polish. The slightly acid flower buds are used in salads. In Britain, in cultivation, the Judas tree usually flowers after the first leaves have appeared.

ANAGYRIS

A. foetida L. **49**   Bean Trefoil

An obnoxious-smelling shrub with trifoliate deciduous leaves and clusters of yellowish laburnum-like flowers. A spreading bush, 1–3 m. high, with young green stems covered in whitish hairs. Leaves stalked, with 3 narrow-elliptic leaflets, hairless above and somewhat hairy beneath. Flowers 2½ cm. long, borne in short clusters, yellow, with the standard about ½ the length of the wings and marked with a blackish blotch; calyx bell-shaped with 5 triangular teeth; stamens 10 unfused. Pods hanging, 10–18 cm. by 2 cm., hairless, swollen at one end; seed large, violet. HABITAT: garigue and dry hills; circum-Medit. February–March.

A very poisonous, emetic and purgative plant.

LUPINUS—Lupin

L. angustifolius L. **52**

A dark blue-flowered lupin with noticeably narrow leaflets and with dense spikes of alternately arranged flowers. An annual up to 50 cm. high with erect stems covered in flattened (not spreading) hairs. Leaves hand-shaped; leaflets linear to oblong, blunt, hairless above, with flattened hairs below. Pods with flattened hairs and 4–6 marbled black and white seeds. HABITAT: fields, stony places and cultivated ground; circum-Medit. March–June.

The similar *L. hirsutus L.* (53) has spreading hairs on the stem, shorter spikes of blue flowers in which the upper flowers are arranged in whorls (not alternate). Leaflets much broader, obovate-oblong and with a short point. Pod very hairy with 3–4 brown mottled seeds. HABITAT: cultivated ground, field margins; circum-Medit. March–May.

L. albus L. is distinguished by its white or pale blue-flushed flowers in short spikes. Pods hairy with 2–4 white (unmottled) seeds. Often cultivated for fodder and sometimes naturalized. May–June.

COLUTEA—Bladder Senna

C. arborescens L. **342**   Bladder Senna

A shrub with pale green compound leaves, yellow flowers and unusual barrage-balloon-shaped pods which become dry and papery. An erect spineless shrub, up to 3 m. high; leaves of 7–15 elliptic or obovate leaflets, hairless or with fine hairs below. Flowers 3–8 in erect clusters shorter than the leaves, yellow, sometimes with reddish markings, about 2 cm. long; calyx with flattened whitish hairs. Pods large, 6–7 cm. by 3–4 cm., blown up and parchment-like, veined and usually hairless, not splitting when ripe. HABITAT: woods, bushy places and calcareous hills; circum-Medit. April–July.

The leaves have been used as a substitute for senna; the seeds contain an alkaloid, cytisine, which is poisonous to cattle.

CYTISUS—Broom

C. triflorus L'Hérit. **51**
An erect twiggy shrub with trefoil leaves and terminal leafy clusters of large yellow flowers stained with reddish brown. Plant 1–2½ m. high, with elongated hairy branches. Leaves large, leaflets elliptic to obovate, hairy and ending in a fine point, the central leaflet largest. Flowers 1–3 in the axils of upper leaves, with an upturned yellow standard, stained and lined with reddish brown, and shorter than the keel. Pod with dense flattened hairs, 3–3½ cm. by ½ cm. Leaves and pods go black on drying. HABITAT: woods and thickets, common on the littoral; W. Medit. to Greece, Morocco, Tunisia. April–May.

CALYCOTOME—Thorny Broom

C. infesta (Presl.) Guss. (C. spinosa (L.) Link) **55** THORNY BROOM
An extremely spiny broom-like shrub with numerous yellow flowers making a fine show of colour in the maquis in spring. A densely branched shrub, 1–2 m. high, with small trefoil leaves of obovate leaflets, and lateral shoots reduced to spines; branches of young stems smooth, hairless. Flowers in a group of 2–4, or solitary, with flower stalks 2–3 times as long as the calyx. HABITAT: sunny slopes and rocky places on acid soils; W. Medit. from Spain to Italy, Algeria, January–May.

The similar *C. villosa (Poir.) Link* has flower clusters of 6–15 flowers with stalks of clusters the same length or longer than the calyx. Pods very hairy. Branches furrowed and lightly covered with grey hairs; leaflets silvery-hairy below. HABITAT: rocky hillsides and bushy places; largely E. Medit., from Corsica to Palestine, Algeria. March–June.

GENISTA—Broom

G. cinerea (Vill.) DÇ. **57**
A greyish-white-stemmed spineless broom-like shrub 1–2 m. tall with small hairy yellow flowers ranged in clusters along the stem. Stems densely branched, erect and finally arching, cylindrical, ribbed; leaves narrow lance-shaped, ½ cm. long, absent on old branches. Flowers in erect clusters of 1–5; flowers short, 12 mm. long, with hairy keel and wings pressed to

the standard; calyx hairy; pod 15–18 mm. long. HABITAT: rocks, edges of woods on calcareous soil; W. Medit. from Spain to Italy; Algeria, Tunisia. March–June.

The branches are used for broom making.

G. equisetiformis Spach. 58

A low shrub forming rounded erect rush-like and almost leafless bushes with terminal rounded heads of yellow flowers. Stem green, ribbed, about 60 cm. high, with scale-like leaves 2 mm. long ranged alternately along the stem; lateral branches thread-like. Flower heads spherical at the ends of the branches, 3 cm. across and about 20-flowered; flowers 1½ cm. long with densely hairy calyx and petals; keel and standard of similar length. HABITAT: hot dry places; Spain. Spring.

G. hirsuta Vahl. 59

A gorse-like spiny shrub with dense silvery pyramidal spikes of yellow flowers. Densely branched with many small lateral branches ending in fine sharp spines; young branches densely hairy. Leaves lance-shaped, 1 cm. long. Flowers borne laterally in compact conical spikes clustered together at the ends of the branches. Flower spikes, calyx and petals with long white silky hairs; flowers 1¼ cm. long, keel twice the length of standard. HABITAT: bushy places, sandy uncultivated ground; W. Medit., Spain, Morocco. March–May.

G. acanthoclados DC. 60

A low much-branched spiny shrub with narrow trefoil leaflets and yellow broom-like flowers with outer petals and calyx covered with silky hairs. A variable plant. Shrub ½–1 m. high with much branched greyish stems with fine longitudinal furrows, ending in sharp erect spines. Leaves almost stalkless, of 3 narrow linear leaflets folded lengthwise. Flowers, 1 cm. long, on spiny terminal branches. Pods egg-shaped, silky, 1–2-seeded. HABITAT: dry hills and rocks; Greece to Turkey. June–July.

RETAMA

R. monosperma (L.) Boiss.

A shrub or small tree with rush-like leafless branches and little clusters of white flowers ranged along them. Up to 3 m. high with ribbed, flexible branches; leaves at first lance-shaped, later trifoliate. Flowers 1–1½ cm. long, white, hairy. A very variable plant. HABITAT: maritime sands, and low hills particularly in loose soil and screes; W. Medit., Spain, Morocco, Algeria. Early spring to May.

SPARTIUM

S. junceum L. 56   SPANISH BROOM

An untidy stiff rush-like bush bearing large yellow fragrant flowers. Shrub 1–3 m. high, with slender, erect, almost leafless greyish-green branches, cylindrical in section. Leaves small, simple, lance-shaped, soon falling. Flowers 2 cm. long, with a conspicuous and rounded standard, borne in stiff terminal spikes. Pods hairless, 6–8 cm. by 7 mm., black when ripe. HABITAT: a plant of the maquis, dry slopes and dry woods, largely on calcareous soils; circum-Medit. May–August.

The flowers produce a yellow dye; the stems are used for basket making and yield a fibre used for weaving. The plant contains an alkaloid which is a purgative, emetic and diuretic.

ONONIS—Restharrow

O. natrix L. 343

A showy perennial with very sticky leaves and large yellow flowers streaked with red. A dense much-branched plant up to ½ m. high, more or less woody at the base. Leaves of 3 narrow egg-shaped and toothed leaflets. Flowers in dense leafy heads at ends of branches, with long flower-stalks, as long as or longer than leaves; flowers solitary and flower-stalk ending in a short point. Flowers 1½ cm. long, petals twice as long as sepals. Pod, 2 cm. long, hairy. HABITAT: sandy and stony places on calcareous soils; circum-Medit. April–July.

O. pubescens L.

A densely hairy annual with rather large yellow pea-like flowers. Stem thick, erect, branched, 10–40 cm. high; leaves and stems covered with long white spreading hairs mixed with glandular hairs. Leaves bright green, undivided towards the base and apex of the stem, trifoliate in the middle. Flowers in dense terminal heads borne singly on stalks shorter than the leaves; flowers about 1½ cm. long, petals as long as calyx, pale yellow, often tinged with red. Pod 7–8 mm. long, seeds smooth. HABITAT: dry hills, stony places and derelict cultivation; circum-Medit. (not Tunisia). April–June.

The similar *O. viscosa L.* has distinctive flower stalks which are longer than the leaves and which end in a long fine point. Leaflets broader egg-shaped; pods larger, 12–15 mm., seeds warted. HABITAT: arid places; circum-Medit. May–June.

O. speciosa Lag. 61

A dense sticky shrub, 1½ m. high, with conspicuous terminal leafless spikes of yellow flowers and neatly toothed trefoil leaves. Leaf stalk 1 cm. long, leaflets oval, 1 cm. by 6 mm., with regular saw-toothed margins, the terminal leaflet stalked and with a stalk ½ the length of the blade. Flower spike dense, bracts heart-shaped, pointed, each subtending 1 or 2 shortly stalked flowers about 12 mm. long; bracts and calyx with shaggy hairs; flowers about 12 mm. long. HABITAT: bushy places; Spain, Morocco. April–June.

O. spinosa L. RESTHARROW

A perennial with pink flowers and spiny branches. Stems stiff, shortly hairy or nearly hairless, branches slender, zigzag and very spiny with spines sometimes in pairs, but usually single; a very variable plant. Lower leaves with 3 leaflets, upper simple; leaflets small, narrow egg-shaped and glandular. Flowers shortly stalked in leafy heads; petals pink, 6–10 mm. long, and one third as long again as sepals. Pods glandular, shorter than sepals; seeds rough, warted. HABITAT: waste and stony places, dry fields; circum-Medit. April–June. British native.

The plant here described is *ssp. antiquorum (L.) Briq.* which seems to be the commonest on the Mediterranean littoral.

MEDICAGO—Medick

M. arborea L. **64** (Fig. IVf)   TREE MEDICK
A small ornamental shrub with grey woody stems and white silky hairs on the young branches, bearing heads of bright golden-yellow pea-like flowers, and later the characteristic corkscrew-shaped fruits of the genus. A dense shrub 1–3 m. high with trefoil leaves; leaflets obovate with wedge-shaped bases, shallowly toothed apex and silky hairs on the lower surface. Flowers 1 cm. long, borne in rather dense clusters of about 12. Pods flattened into a single spiral, 1–1½ cm. across with a hole in the centre; surface of pod netted. HABITAT: rocky places; native of Greece and the Islands, W. Turkey. April–June. Often grown as an ornamental plant and sometimes occurring as a casual.

M. marina L. **62** (Fig. IVa)   SEA MEDICK
A densely white-woolly plant with rounded heads of bright pale-yellow flowers, found growing in sandy places by the sea. It has the corkscrew-shaped pods and the trefoil leaves of the genus. A spreading perennial plant with obovate to egg-shaped leaflets almost hidden in white woolly hairs. Flowers small, 6–8 mm. long in rounded heads at the ends of the branches. Pod of about 3 coils with or without spines, and woolly like the rest of the plant. HABITAT: maritime sands; circum-Medit. April–June.

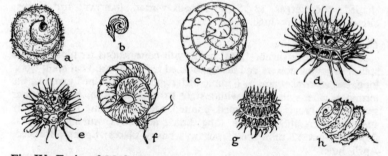

Fig. IV. Fruits of *Medicago spp.*
a *M. marina* (×1½) **b** *M. lupulina* (×1½) **c** *M. orbicularis* (×1) **d** *M. polymorpha* (×2½) **e** *M. minima* (×2) **f** *M. arborea* (×1) **g** *M. arabica* (×1) **h** *M. rigidula* (×2)

M. lupulina L. (Fig. IVb)   BLACK MEDICK

A spreading annual or short-lived perennial, 10–30 cm. high, with rounded heads of numerous tiny yellow flowers and tiny kidney-shaped fruits which turn black when ripe. Flower heads 3–8 mm. long on stalks longer than the leaf stalks; flowers 2–3 mm. long. Pods 2 mm. in diameter, coiled almost in a complete circle, 1-seeded. HABITAT: grassy places, track sides; circum-Medit. April–May. British native.

M. orbicularis (L.) Bartal. (Fig. IVc) is distinguished by its smooth flattened disk-like pods, 15 mm. in diameter, formed of 3–6 tight coils. Flowers in clusters of 1–3 on stalks shorter than the leaf; stipules cut into narrow segments. A spreading practically hairless annual 20–70 cm. long. HABITAT: cultivated ground, olive orchards; circum-Medit. March–June.

M. polymorpha L. (M. hispida Gaertn., M. denticulata Willd.) (344, Fig. IVd) has spiny pods of 2–6 coils with their surfaces strongly netted with veins. Pods 4–8 mm. in diameter, barrel shaped and flattened at each end, and each coil has 2 rows of diverging long or short, hooked or curved spines. Flowers in clusters of 3–8, each flower 5 mm. long. A spreading annual, 5–60 cm. HABITAT: rocky places, track sides, fields; circum-Medit. April–May. British native.

M. rigidula (L.) Desr. (Fig. IVh) has more or less spherical pods which are covered with flattened woolly hairs or glandular hairs, of 4–7 coils, 7–9 mm. in diameter. Spines on coils conical, hooked, grooved at the base and unevenly spaced. Plant covered with flattened hairs; 20–60 cm. long. HABITAT: fields, banks and track sides; Spain to Palestine, Egypt, Algeria, Morocco. April–May.

M. arabica (L.) Huds., SPOTTED MEDICK (Fig. IVg), has a more or less spherical pod of 4–7 coils, 4–6 mm. in diameter, and with a double row of widely spreading hooked or curved spines. A sparsely hairy spreading annual 10–60 cm. long with large leaves and leaflets up to 2½ cm. long usually with a dark blotch in the centre. Flower heads 1–5, flowers 4–5 mm. long. HABITAT: track sides, fields and cultivated places; circum-Medit. March–June. British native.

M. minima (L.) Bartal., SMALL MEDICK (Fig. IVe), has more or less spherical pods 4–6 mm. in diameter, of 4–5 coils, with many long close-set spreading spines hooked at the tips. Flowers in clusters of 1–8, flowers 4 mm. long. A densely hairy spreading or erect annual 5–20 cm. long with egg-shaped to lance-shaped uncut or shallowly toothed stipules. HABITAT: dry places, fields, banks and track sides; circum-Medit. March–June. British native, East Anglia.

TRIGONELLA

T. balansae Boiss. & Reut. (Fig. Va)

A robust yellow-flowered annual with spherical heads of drooping flowers and large sickle-shaped pods. A conspicuous plant with pale green leaves; stems up to 50 cm. long, either erect or spreading. Leaves trefoil with obovate, toothed and stalked leaflets 2 cm. long. Flowers bright yellow, 1 cm. long, about 14 in each long-stalked flower head. Pods conspicuous, 2½ cm. by 4 mm., curved in a semi-circle and spreading outwards from the flower stalk. HABITAT: in cultivated ground and sandy places; Greece and the Islands, Turkey. March–May.

The similar *T. graeca Boiss. & Sprun.* has deeper orange flowers arranged in an elongated spike, and pods which are flattened, circular in outline with conspicuous broad papery wings and widely spaced veins. HABITAT: stony hills; Greece and the Islands near the mainland. March–May.

T. coerulescens (Bieb.) Hal.

A small clover-like plant with dense rounded heads of bluish flowers, which is sometimes abundant on the low hills near the sea in Greece. A hairy spreading annual, up to 20 cm. high, with softly hairy stems and leaves. Leaves trefoil with wedge-shaped obovate leaflets, toothed at the apex. Flower-stalk long, carrying the flower head well above the leaves; flower head at first egg-shaped but becoming oblong in fruit. Flowers bluish white; sepals softly hairy with long pointed tips. Pods spear-shaped, hairy, longer than calyx. HABITAT: dry hills; Dalmatia to Syria. March–April.

MELILOTUS—Melilot

M. indica (L.) All. (Fig. Vb)

Probably the most widespread of the 10 or so species of Melilots that occur commonly in the Mediterranean region. Distinguished by its minute yellow flowers in narrow spikes and its minute seed pods. An annual, 10–40 cm. high, with erect branched and hairless stems. Leaves trefoil,

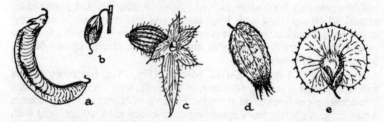

Fig. V. Fruits of *Leguminosae*
a *Trigonella balansae* (× 1) **b** *Melilotus indica* (× 4) **c** *Trifolium clypeatum* (× 1½) **d** *Physanthyllis tetraphylla* (× 1½) **e** *Hymenocarpus circinnatus* (× 1)

leaflets obovate to narrow lance-shaped, toothed. Flowers very small, 2–3 mm. long, in long narrow spikes longer than the leaves and elongating in fruit. Pods globular, 2–3 mm. long, stalkless, hanging, whitish-grey. HABITAT: damp sandy places, fields; circum-Medit. April–June. British casual.

The only common white-flowered Melilot, *M. alba Medic.*, is distinctive. It is a tall slender biennial up to 1½ m. high. Flowers in slender, lax clusters; flowers and pods 4–5 mm. long, the latter brown when ripe and with a netted surface. HABITAT: uncultivated, dry places, fields and track sides; European and Asiatic Medit., Egypt. May–August. Naturalized in S. England.

TRIFOLIUM—Clover, Trefoil

There are over 60 species of clovers and trefoils in the Aegean alone and many species are common over the whole of the Mediterranean region. A few species are distinct and can be easily recognized, but the majority require careful study.

T. uniflorum L.   63

This plant forms a low spreading mat with white or pinkish-purple, usually solitary clover-like flowers carried only a little way above the leaves. Leaflets egg-shaped or broader, with adpressed hairs; stipules broadly egg-shaped at the base and long-pointed. Flower stalk very short, shorter than stipules, 1–3 flowered; flowers 1½ cm. long, with a conspicuous elongated standard more than twice the length of the calyx. HABITAT: dry hills and on the littoral. France to Turkey. March–June.

T. tomentosum L.   WOOLLY TREFOIL

Easily recognized by its spherical woolly-white flowering and fruiting heads. Stems hairless, spreading, 5–20 cm. Leaves with broadly obovate toothed leaflets. Flower-stalks very short or absent so that the heads appear closely pressed to the stem in the axils of the upper leaves. Flowers in small pink heads, ½ cm. across, which enlarge in fruit to 1–1½ cm. across; fruit heads hidden by woolly white hairs. HABITAT: stony places, track sides and cultivated ground; circum-Medit. March–May. A British casual.

T. clypeatum L.   (Fig. Vc)   SHIELD CLOVER

A large showy pinkish-white-flowered clover with broad green leaves and comparatively large flowers. The sepals enlarge very much in fruit and become increasingly more conspicuous as the teeth spread outwards in the form of a shield. A coarse, spreading, rather weak-stemmed, branched annual, 30–50 cm. high, with large leaves and broad obovate toothed leaflets. Flower heads solitary, egg-shaped; flowers 2½ cm. long. Calyx teeth of 5 light green leafy lobes, with the lowest lobe much longer than the rest. HABITAT: hills and grassy places in the maquis; E. Medit., Greece to Palestine. March–April.

T. stellatum L.   65   STAR CLOVER

The star clover is not easily mistaken for, after flowering, the globular fruiting heads become starry as the sepals dry out and spread open and turn a dark crimson colour, often with a white centre. A hairy annual,

99

5–25 cm. high, covered with soft spreading white hairs, with trefoil leaves of obovate toothed leaflets. Flower head spherical, appearing silky with the long hairs on the calyx teeth. Flowers pale pink, about the same length as the long-pointed calyx teeth, which are at first parallel, and then spread outwards into a star. HABITAT: dry places, sands, track sides and fields; circum-Medit. March–June. Naturalized in Sussex.

T. purpureum Lois. 66   PURPLE CLOVER

A robust clover with narrow leaflets and very large conical or cylindrical heads of bright rosy-purple flowers. A striking and attractive species, which is very variable. An annual up to 60 cm. high, erect and branched with rather sparse spreading hairs on stems and leaves, with trefoil leaves, the leaflets all narrow lance-shaped. Flower heads terminal, long-stalked; the flowers at the base of the head open much before those at the apex. Flowers with petals much longer than the calyx, which has narrow hair-like teeth covered in long spreading stiff white hairs. HABITAT: fields and sandy dry places; European and Asiatic Medit., Egypt. March–July. Often cultivated.

T. pilulare Boiss.

A quite unmistakable trefoil with tiny flower heads looking like miniature balls of cotton wool spread over the surface of the ground. The heads are less than 1 cm. across and are covered in very delicate fluffy white hairs. A flat-growing annual spreading to 40 cm. with trefoil leaves and hairy obcordate leaflets. Flowers inconspicuous, white, hidden in the woolly hairs of the calyx teeth; fruiting heads spherical, becoming increasingly 'woolly' as teeth of calyx elongate and become curved and wavy. HABITAT: stony hills and shady places; E. Medit., Dodecanese, Turkey to Palestine. April–May.

ANTHYLLIS—Kidney-Vetch

A. vulneraria L. 67   KIDNEY-VETCH, LADIES' FINGERS

A very variable plant with a number of well-recognized sub-species. The most widespread has yellow flowers, but the commonest form along the Mediterranean is probably *ssp. spruneri* (*Boiss.*) *Hay.* (**67**) with deep rose or crimson flowers and a conspicuous creamy white calyx tipped with purple. An annual or perennial with spreading or erect hairy stems ending in large flower heads. Leaves very variable, basal leaves undivided and ovate, or compound with several pairs of leaflets and a terminal much larger leaflet; stem leaves few but with leaflets more equal in size. Flower heads solitary or often paired, dense, and encircled by a conspicuous narrow-lobed leafy bract; calyx inflated, covered with silky hairs, teeth unequal. HABITAT: dry fields, meadows and hill sides; W. Medit. to Turkey, N. Africa. April–August. The yellow form is a British native.

A. cytisoides L. 69

A dense rounded shrub with greyish white stems and leaves, and lax spikes of pale yellow flowers half hidden by broad leafy bracts. Stems erect, 30–80 cm., spineless and covered in velvety white hairs; lower leaves undivided, up to 3 cm. by 1 cm., upper trefoil. Flowers in clusters of 2–5 in axils of broadly egg-shaped pointed bracts; flowers small, 8 mm. long; calyx with white woolly hairs. HABITAT: exposed places by the sea, dry banks; Spain to France, Morocco, Algeria. April–June.

A. barba-jovis L.

A shrub with silver-white compound leaves and dense terminal heads of pale yellow flowers. Stems erect, silvery white, up to 1 m. high. Leaves of 9–19 narrow lance-shaped leaflets. Flowers numerous in stalked, flattened heads; calyx silky. HABITAT: rocks by the sea; France, Corsica, Italy, Tunisia, Algeria. April–June.

PHYSANTHYLLIS

P. tetraphylla (L.) Boiss.   70   (Fig. Vd)

A pale grey-green silky-hairy annual with rounded leaflets and pale yellow flowers touched with orange, and much swollen hairy calyx. Stems spreading over the ground with few branches; leaves compound with the terminal leaflet orbicular or obovate and much larger than the unequal lower leaflets. Heads, 3–6 flowered, stalkless in the axils of the leaves. Calyx very conspicuous and covered in dense silky hairs, becoming swollen-ellipsoid in fruit and often turning reddish. HABITAT: in cultivated ground, olive groves, fields; circum-Medit. March–July.

HYMENOCARPUS

H. circinnatus (L.) Savi   (Fig. Ve)

A plant with small yellow flowers, distinctive compound leaves, and very unusual looking fruits. The leaves have disproportionately large egg-shaped terminal leaflets, and the fruits are flattened disks with wavy spiny margins. A prostrate or ascending annual up to 50 cm. with simple lower leaves and upper of 2–4 pairs of small leaflets and a large terminal leaflet, 3–4 cm. by 2 cm. Flower heads long-stalked, of 2–8 small yellow flowers. Calyx hairy, bell-shaped not inflated, with 5 long teeth. Fruits almost circular, about 1½–2 cm. across, hairy and netted. HABITAT: cultivated ground, turf and dry hills; circum-Medit. (not Morocco?). March–May.

BISERRULA

B. pelecinus L.   (Fig. VIb)

A spreading hairy annual with compound leaves, inconspicuous heads of small whitish flowers, and unmistakable fruits reminiscent of a 2-edged saw. Leaves of 7–15 pairs of small leaflets. Flower heads short-stalked, of 3–10 flowers, white, tipped with blue; calyx hairy. Fruit 1½–3 cm. by 6–8 mm., flattened and strongly toothed on each margin. HABITAT: dry hills, bushy places, track sides; circum-Medit. April–June.

LOTUS

L. creticus L.

A spreading plant with silvery white, very short-stalked trefoil leaves and few-flowered yellow heads. A spreading or ascending perennial, 10–40 cm. long with a woody base and few leaves. Leaflets wedge-shaped, covered in silky hairs, and leaf-stalks much shorter than stipules. Flower stalks 1–8 flowered and much longer than leaves; flowers golden-yellow, about 1½ cm. long; calyx with unequal teeth divided into 2 lips. Pod variable, 3–5 cm. long, cylindrical, straight or slightly curved. A very variable species: other forms are sparsely hairy or hairless, or have golden hairs. HABITAT: dunes, cliffs and maritime sands; circum-Medit. March–June.

TETRAGONOLOBUS

Tetragonolobus purpureus Moench (*Lotus tetragonolobus L.*)   **72, 345**

WINGED PEA, ASPARAGUS PEA

Conspicuous, solitary, dark red pea-like flowers set amongst broad trefoil leaves, and cylindrical pods with 4 broad wavy flanges, distinguish this plant from all others. A softly hairy, spreading annual, 10–40 cm. long, with broad egg-shaped or rhomboidal leaflets up to 2 cm. long. Flowers usually solitary or occasionally paired, 1½ cm. long or more, on stalks little longer than the leaves which subtend them. Pod 2 cm. long with broad undulating wings, as wide as the pod itself. HABITAT: cultivated ground, track sides and banks; circum-Medit. (not Turkey). March–May.
Sometimes cultivated for food.

BONJEANIA

B. hirsuta Reichb. (*Dorycnium hirsutum L.*)

A cottony white plant with conspicuous heads of rather large white flowers flushed with pink and with a short black-purple keel. A spreading hairy perennial with a woody base, 20–50 cm. high. Leaflets 3, with 2 similar stipules, oblong-lance-shaped, 1–1½ cm. by ½ cm., softly hairy. Flowers 5–10, in rather dense heads; flowers 1½ cm. long with very hairy calyx ½ the length of the petals. HABITAT: rocks, stony slopes, sandy places and field verges; European and Asiatic Medit. April–July.

ASTRAGALUS

A. spruneri Boiss.

An attractive plant with heads of rose-purple flowers. A prostrate perennial with compound leaves of 5–12 pairs of egg-shaped acute leaflets with flattened greyish-white hairs. Flower head ovoid, 6 cm. long, of pink flowers flushed with purple; flowers about 2 cm. long on short stalks; calyx with black flattened hairs. Pods 2 cm. long, curved, narrowing to a point, with flattened hairs. HABITAT: dry stony ground; Greece and Aegean Islands, Rhodes. March–May.

This is one example of this enormous genus which is spread throughout the Mediterranean.

SCORPIURUS

S. subvillosus L. (Fig. VIc)

A plant reminiscent of *Lathyrus pratensis* with small heads of a few yellow flowers, but the leaves and fruits are quite distinct. An annual up to 10–50 cm. high with broadly lance-shaped single-bladed leaves narrowed at the base into a long stalk. Flower heads of 1–4 flowers on very long stalks, longer than the leaves; flowers about 8 mm. long; calyx bell-shaped, hairy, with teeth longer than the tube. Fruits twisted and contorted into an irregular corkscrew with the outer sides covered with straight, crooked or divided bristly spines. HABITAT: cultivated ground and dry places; circum-Medit. March–June.

The similar *S. sulcatus L.* is a hairy plant with a tighter and more regular spiralled fruit, and with flowering stems at first shorter than the leaves; fruit with sharp slender spines; calyx teeth shorter than tube. HABITAT: cultivated bushy places; circum-Medit. March–June.

The similar *S. vermiculatus L.* usually has solitary flowers on stalks at first shorter than the leaves. Fruit covered on the sides with short flattened swellings. HABITAT: fields; European Mediterranean, N. Africa. March–April.

PSORALEA

P. bituminosa L. 73   PITCH TREFOIL

A robust clover-like plant with rounded heads of dull blue-violet flowers; when crushed, the whole plant gives off an unusual smell reminiscent of tar. An erect dark green, sparsely branched perennial, 50–100 cm. high. Leaves trefoil, the lower with elliptic or egg-shaped leaflets, the upper narrower and more pointed; all leaflets sparsely hairy and glandular. Flower stalks very long and stiff bearing 10–15 flowers, with toothed bract closely investing the globular heads. Calyx hairy, with long pointed teeth, little shorter than the narrow flowers. HABITAT: dry and arid places, by roadsides and thickets; circum-Medit. April–July.

ROBINIA—Locust

R. pseudoacacia L.   'ACACIA' TREE, FALSE ACACIA

A native tree of North America which has been widely planted in the Mediterranean region. It is a medium-sized tree with deeply furrowed bark, spiny stems, pale green foliage, and sweet scented drooping clusters of white flowers. Leaves compound, of 7–15 oval leaflets 2½–5 cm. long, with 2 strong spines at the base of the leaf-stalk. Flowers white, about 2½ cm. long, in a hanging cluster 12–15 cm. long. Pods reddish-brown, 8 cm.

long, pendulous. HABITAT: widely planted along roads and railway banks, where it gives shade and helps to consolidate the soil; sometimes naturalized and often suckering; throughout Medit. May–June.

The wood lasts well in the soil and is used for gate posts.

CORONILLA

C. emerus L. ssp. **emeroides** (*Boiss. & Spr.*) *Hayek* (**C. emeroides** *Boiss. & Spr.*) 68 (Fig. VIe) SCORPION SENNA

A dense symmetrical shrub with green branches, glossy dark green leaves and clusters of rather large yellow flowers. Up to 2 m. high; leaflets obovate of 2–4 pairs with a slightly larger terminal leaflet. Flowers yellow, often tipped with red. Flowers 1½–2 cm. long, in a head of 2–7 flowers; the elongated bases of petals are widely separated and much longer than calyx. Pods hanging, 5–10 cm. long, straight and jointed into 7–10 sections. HABITAT: woods and hills, always on limestone; European and Asiatic Medit. March–May.

The bitter foliage is used to adulterate true senna (Cassia acutifolia, Delile, *and other species*). *Cultivated in Britain since the sixteenth century.*

C. scorpioides (*L.*) *Koch* 346

A blue-grey (glaucous) hairless annual with smooth rather thick stalkless rounded leaves, small yellow flowers, and long, curved, very narrow pods reminiscent of a scorpion's tail. An erect much branched annual up to 40 cm. high. Leaves usually with 3 leaflets with the central leaflet very large, broadly egg-shaped, blunt and the 2 lateral leaflets a ¼ the size and almost circular. Flowers small, ½ cm. long, 2–4 on the end of a long flower stalk. Pods, 5–7 cm. by 1–2 mm., constricted between seeds and curved like a sickle, with a fine bristle at the end. HABITAT: fields, waste places and cultivated ground; circum-Medit. March–June.

C. juncea L. 71

A sparsely branched shrub with straight grey-green rush-like stems and long-stalked heads of bright yellow flowers. Stems up to 1 m. high, rather pithy and easily squashed; leaves with 3–7 narrow leaflets. Flower heads on stalks longer than the leaves, 2–8 flowered in a dense umbel; pod long, narrow, 4-angled, slightly curved and jointed. HABITAT: hillsides, dry woods and roadsides; W. Medit., Spain to Dalmatia, Morocco to Tunisia. March–June.

HIPPOCREPIS—Horse-shoe Vetch

H. unisiliquosa L. 347 (Fig. VIa)

A small yellow vetch with striking pods looking like a string of horseshoes placed side by side. A hairless spreading annual, 20–30 cm. long, with compound leaves with 4–7 pairs of narrow-oblong notched leaflets and a terminal leaflet. Flowers stalkless or very short-stalked, solitary or paired in the axils of the leaves; petals ½ cm. long. Pod stalkless and slightly

curved, 3 cm. by ½ cm., with 7–10 deep rounded horseshoe-like cavities.
HABITAT: maquis, garigue, edges of thickets; circum-Medit. March–May.

The similar *H. multisiliquosa L.* is distinguished by its much longer
flower stalks which are nearly as long as the leaves, and the clusters of
2–6 flowers. Pod similar but strongly curved, often glandular and reddish.
HABITAT: sunny, dry hills; circum-Medit. March–May.

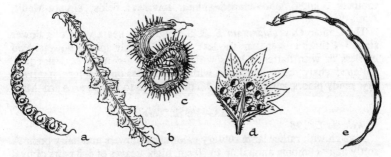

Fig. VI. Fruits of *Leguminosae*
a *Hippocrepis unisiliquosa* (×¾) **b** *Biserrula pelecinus* (×1½) **c** *Scorpiurus
subvillosus* (×2) **d** *Onobrychis aequidentata* (×2) **e** *Coronilla emeroides*
(×1)

HEDYSARUM

H. coronarium L. **74**

A sainfoin-like plant with a dense head of large beautiful carmine
flowers and compound leaves. A herbaceous perennial with ascending
stems, 20–100 cm. high; leaves large with 2–7 pairs of large oval leaflets up
to 3¼ cm. by 1½ cm. and a terminal leaflet. Flower heads oblong, 6 cm. by
3 cm., on stalks as long as or longer than the leaves; flowers nearly 2 cm.
long. Pod elongated, flattened, with 2–4 rounded joints, hairless but with
sharp rough projections. HABITAT: fields; Spain to Turkey; N. Africa.
April–May. Cultivated and naturalized.

H. capitatum Desf. has fewer rose coloured flowers (4–10) in an umbel;
much smaller and narrower leaflets (1–5 mm. across), and fruit covered
with short dense hairs and hooked spines. HABITAT: dry bushy places,
clayey hills; Spain to Greece, Algeria. April–May.

ONOBRYCHIS

O. caput-galli (*L.*) *Lam.* **348**

An annual with greyish hairy leaves and narrow spikes of very small
pink flowers, and unusual-looking hard spiny fruits. An erect or spreading
plant, 20–40 cm. high, with 5–7 pairs of narrow-oblong hairy leaflets.
Flowers about 6 mm. long, in close heads of 3–7 flowers, flower heads long
stalked; calyx teeth long, pointed, hairy, twice length of tube. Pods about
8 mm. in diameter, somewhat flattened, densely spiny and indented with

depressions on each face and with a coxcomb-like crest of unequal flattened spines round the margin; pod 1-seeded. HABITAT: arid places, track sides, common on the littoral; circum-Medit. April–May.

The similar *O. crista-galli L.* has very narrow leaflets, a flower stalk shorter than the leaves, bearing 2–3 bright pink flowers, and calyx teeth 4–5 times as long as the calyx tube. Pod 3-seeded, and crest of pod of 3–4 toothed, pointed lance-shaped spines. HABITAT: fields; circum-Medit. April–June.

The similar *O. aequidentata S. & S.* (Fig. VId) has a very long flower stalk 3–4 times longer than the leaf, and 1–3 widely spaced flowers. Pod hairless or with flattened hairs with a crest of 5–6 more or less equal, flattened, sharp teeth, sometimes with a few spines on each face. HABITAT: dry, sandy places, sunny hills; E. Medit., France to Palestine. April–May.

VICIA—Vetch, Tare
V. hybrida L. 75
A vetch with rather large solitary pale yellow flowers and hairy pods. A softly hairy climbing annual up to 50 cm. high. Leaves of 4–8 pairs of oval flat-topped leaflets and a terminal branched tendril. Flower practically stalkless in the axil of the leaf, 2½ cm. long, lemon-yellow and with the upper petal (standard) densely hairy on the outside, often veined with purple. Pod yellowish, covered with long spreading hairs. HABITAT: fields, track sides, cultivated ground; European and Asiatic Medit., Algeria. March–June.

V. lutea L. is very similar but the upper petal is hairless on the outside; the leaflets are broadly linear and blunt-ended with a fine point. Flowers sulphur yellow and pods hairy with swellings at the base of the hairs. HABITAT: similar to *V. hybrida*, but circum-Medit. A British native.

V. dasycarpa Ten. 349
This plant is similar to the Tufted Vetch, *Vicia cracca*. Flowers violet-purple, sometimes with a reddish base, clustered into conspicuous heads. A rather robust scrambling biennial with flattened hairs, and leaves of 5–9 pairs and a terminal branched tendril; stipules half-arrow-shaped. Flower clusters rather dense on stalks longer than the leaves, flowers 6–12, at first horizontal and later drooping. Petals 1½–2 cm. and 3–4 times longer than the calyx. Pod usually hairless, 4 cm. by 10–12 mm. HABITAT: hedges, stony and bushy places, fallow land; circum-Medit. March–June.

The similar *V. villosa Roth* is a softly hairy plant with spreading hairs and blue-violet flowers. Flower clusters plume-like before flowering owing to the long hairy teeth on the calyx; flowers pendulous. HABITAT: fields, bushy places and cultivated places; European and Asiatic Medit., Morocco. May–June.

PISUM
P. elatius Stev. 76   WILD PEA
A hairless blue-green tendril-climbing plant with compound leaves and

rose-purple flowers with dark blackish-purple wings. An annual climbing 1 m. or more; leaves of 2–3 pairs of oval leaflets, and conspicuous half-heart-shaped stipules shorter than the leaf stalks, with terminal branched tendrils. Flowers large, 2–3 cm. across, 2–3 on a long stalk; pod up to 1½ cm. broad. HABITAT: thickets and woods; circum-Medit. April–June.

LATHYRUS

L. aphaca L. 77   YELLOW VETCHLING
   A famous plant among botanists for the leaves are modified to long tendrils and the stipules are broad and leaf-like. A delicate scrambling hairless rather grey-green annual, up to 50 cm. or more long, with paired, broadly triangular stipules joined at the base and clasping the stem, and with a much longer central whip-like tendril. Flowers usually solitary, on stalks longer than stipules, yellow, 8–15 mm. long. Pods 2–3 cm. by 6 mm., curved, with 4–5 seeds. HABITAT: a plant of limestone soils, dry places and fields; circum-Medit. March–June. British native.

L. clymenum L.
   An annual with hairless climbing stems and clusters of 2–4 flowers with conspicuous purple standards and bluish wings. A very variable plant. Lower leaves simple lance-shaped, without tendrils, upper with 2–4 pairs of narrow (3–11 mm.) leaflets and a terminal branched tendril; leaf stalks with narrow wings. Flowers quite large, 15–20 cm. long, on a flower stalk as long as or longer than the leaves. Pods flattened, channelled along the back. *Var. articulatus (L.) Arcang.* has larger rich purple-madder flowers with white or rose coloured wings and 1–2-flowered heads. A narrower-leaved (leaflets 2–5 mm. broad), more bluish-grey plant. HABITAT: sandy and stony places, and borders of fields; European Medit., Crete, Asia Minor, N. Africa, March–June.
   The rather similar *L. tingitanus L.*, the TANGIER PEA (78), has much larger, showy, scentless, carmine-purple flowers with a circular standard 2 cm. or more in diameter. A hairless plant; pod shining when mature. HABITAT: open woods, thickets, screes; W. Medit.; Spain, Morocco, Algeria. April–July. Often cultivated and naturalized.

L. ochrus (L.) DC.
   An unusual bluish-grey annual with large narrow egg-shaped 'leaves' (actually wings of the leaf stalk) and 3 short tendrils at the apex of the blades. Flowers pale yellow, on short stalks among the upper 'leaves'. The lower and middle 'leaves' are entire, 5 cm. by 1½–2 cm., with the base running down the stem in a flange, and 3 tendrils at the apex. Upper 'leaves' have 1–2 pairs of oval leaflets and terminal branched tendrils. Flowers 16–18 mm. long; pods 5 cm. long with 2 wings on the upper margin. HABITAT: cornfields and dry places; circum-Medit. March–June.

L. cicera L. 350
   Easily distinguished by its rather showy solitary brick-red or crimson

flowers, and narrow paired pointed leaflets. A hairless climbing plant up to 60 cm. with a narrowly winged stem. Leaves shortly stalked with 2 long narrow, pointed leaflets, 7–8 cm. by 2 mm., and a terminal branched tendril; stipules half-arrow-shaped. Flowers 10–15 mm. long, borne singly in the axils of the upper leaves on a stalk longer than the leaf-stalk but shorter than the leaves. Pod hairless, 3–4 cm. by 8–10 mm. HABITAT: fields and cultivated places; circum-Medit. March–June.

OXALIDACEAE—Wood Sorrel Family

Herbaceous plants, often with a fleshy rootstock and in our species with a sour juice. Leaves compound, with 3 leaflets, often folding up at night, stipules usually absent. Flowers sometimes solitary, but commonly in an umbel; petals and sepals 5; stamens 10, fused at the base (sometimes 5 sterile); ovary 5-celled and with 5 free styles; fruit a capsule. A family largely of the tropics and sub-tropics; some are cultivated for ornament. *Oxalis* is the only Mediterranean genus.

OXALIS—Wood Sorrel

O. pes-caprae L. (O. cernua Thunb.) **79**

BERMUDA BUTTERCUP, CAPE SORREL

An attractive and conspicuous butter-yellow-flowered plant with long stalked, bright green trefoil leaves. A stemless herbaceous perennial with a rootstock forming many bulbils the size of peas at the surface of the soil. Leaves somewhat fleshy; leaflets heart-shaped. Flowers in an umbel of 6–12, buds drooping; petals 2–2½ cm. long, occasionally double. HABITAT: cultivated ground, orchards, walls; circum-Medit. February–April. A casual in W. Britain.

This plant has become a serious weed in many Mediterranean orange groves and orchards. It is a native plant of the Cape of Good Hope, and was brought to the Mediterranean at the end of the eighteenth century. It spreads very rapidly by bulbils which are produced in great numbers.

O. articulata Savigny (O. floribunda Lehm.) **81**

This pink-flowered 'wood sorrel' is a native of Brazil which has become naturalized in some parts of the Mediterranean littoral. A perennial with short, thick, swollen rootstock, and with all leaves and flowering stems arising from it; leaflets hairy, with pale orange spots beneath. Flowers 3–16, on a long-stalked umbel; petals 1–1½ cm. long, purplish-pink. No bulbils are formed in this species. HABITAT: cultivated ground. June–December. A casual in S.W. Britain.

GERANIACEAE—Geranium Family

Herbs or undershrubs with lobed or compound leaves and usually with paired stipules. Flowers often showy, solitary or in spreading heads. Sepals usually 5, free or fused to the middle; petals 5; stamens 2–3 times the number of petals, more or less fused at the base. Ovary 3–5-celled, ending in a long beak formed by the fused styles. Fruit in our genera breaking into 1-seeded sections attached to the style which is elastic. Some species are ornamental, notably the garden geraniums, *Pelargonium spp.*

GERANIUM: lobes of leaf as broad as long; all stamens fertile. Seeds dispersed by style rolling elastically upwards.

ERODIUM: leaves usually compound with leaflets longer than broad; 5 stamens infertile, without anthers. Fruit dispersed by corkscrew-like twisting of the style.

GERANIUM—Cranesbill

G. molle L. 82   DOVE'S-FOOT CRANESBILL
A hairy annual with small pinkish flowers, rounded cut leaves and beaked fruit. A branched, rather weak, often spreading plant, to 40 cm. or more. Leaf blades cut to more than half their width into narrow lobes. Petals rosy-purple with a deep notch in each, usually only 4–4½ mm. long, but in *forma grandiflorum Lge.* (82), which is common in Greece, they may be 5–6 mm. long. Fruit with a beak-like style 1–1½ cm. long, with wrinkled and hairless carpels; seeds smooth. HABITAT: track sides, dry hills, grassy places and hedges; circum-Medit. March–June. British native.
The similar ROUND-LEAVED CRANESBILL, *G. rotundifolium L.* (352), has leaves characteristically circular in outline with rounded lobes cut to half the width of the blade. Flowers pink, ½ cm. across, inconspicuous, with rounded petals. Fruit beaked, with smooth but glandular hairy carpels; seeds pitted. HABITAT: cultivated ground, waste places, track sides, walls and burnt ground; circum-Medit. March–June. A British native.

G. tuberosum L. 83   TUBEROUS CRANESBILL
A delicate cranesbill with a distinctive globular, underground, nut-like tuber; with long-stemmed root leaves and slender flower stems which are leafless below. Stems 20–40 cm. high, finely hairy, branched above. Leaves mostly from the roots, cut to the centre into 5–7 narrow segments which are further divided. Flowers about 1½ cm. across, pink-purple with darker veins, and petals, each with a conspicuous rounded notch, twice the length of the hairy sepals. Carpels hairy, not ribbed. HABITAT: cultivated ground, fields, vineyards and meadows; European and Asiatic Medit., N. Africa. March–June.

G. lucidum L.   SHINING CRANESBILL
A delicate pink-flowered annual with rounded shining green leaves which often turn bright red. Up to 30 cm. high but often much shorter, branched from the base. Leaves rather fleshy and brittle with few hairs, long-stalked and almost circular in outline, but deeply cut into five broad blunt lobes which are further divided into rounded lobes. Flower stalk longer than the leaves, 2-flowered; flowers small, 8–9 mm. across, petals

bright pink, not notched at the apex; sepals egg-shaped, long-pointed, forming an angled calyx with transversely ribbed surface. HABITAT: shady rocks, walls, hedges; circum-Medit. March–May. A British native.

ERODIUM—Storksbill

E. gruinum (L.) L'Hérit. 353

A storksbill with medium-sized violet flowers, which fall prematurely, and very long-beaked fruits. An annual with thick, sparsely hairy stems up to ½ m. high. Leaves of 2 kinds, the lower undivided, egg-shaped with heart-shaped bases; the upper divided into 3 triangular acutely toothed leaflets, with the central lobe longer, and itself often 3–5-lobed. Flowers 2–3 in a head on a hairy stalk. Flowers 2 cm. across, petals drop early; calyx enlarging in fruit to 1½ cm. long, elliptic with conspicuous veins and a long fine point. Beak of fruit stout, 8–11 cm. long, carpels covered with stiff bristles. HABITAT: fields and sandy places near the sea; Greece to Palestine; N. Africa, Sicily. February–May.

E. malacoides (L.) Willd.

Distinguished by broad egg-shaped, mallow-like leaves, which are undivided, or divided to less than half the width of the blade into rounded lobes with toothed margins. A glandular-hairy annual, 10–40 cm. high. Flowers rather small, 3–8 in a glandular, sticky umbel standing above the leaves; petals, 4–7 mm. long, lilac, widely spaced, and not touching at their edges. Fruit hairy with 2 glandular pits at the apex, and 2 semi-circular ridges immediately below them, and with a spirally twisted beak, 3–4 cm. long. HABITAT: roadsides and waste places; circum-Medit. February–June.

E. cicutarium (L.) L'Hérit.   COMMON STORKSBILL

A rather delicate, soft, spreading plant with much-cut ferny leaves and small, rose-purple flowers. An annual or biennial up to 60 cm. high, but very variable in size and robustness. Leaves divided into egg-shaped leaflets which are once or twice further dissected; stipules papery and pointed. Flowers in heads of 2–9, petals rosy-purple, often with a blackish spot at the base and usually not touching at their edges; flowers 12–16 mm. across. Beak of fruit 2–4 cm. long. HABITAT: track sides, cultivated ground and sandy uncultivated places; circum-Medit. November–May. A British native.

E. romanum (L.) L'Hérit.

An attractive plant with bright rosy flowers on leafless stalks arising from a basal rosette of leaves. A perennial with a woody rootstock and many stalked leaves which are divided into egg-shaped segments which are further cut into oblong pointed segments. Flowers conspicuous, 1½ cm. across; petals 2–3 times the length of the calyx, egg-shaped and touching at their edges. Beak of fruit 4–5 cm. long. HABITAT: fields; European and Asiatic Medit. March–April.

LINACEAE—Flax Family

Herbs or small shrubs with simple undivided leaves, usually alternate and usually without stipules. Flowers with 4 or 5 sepals, petals and stamens. Stamens usually fused at the base with additional infertile stamens between the fertile ones. Ovary with 3–5 cells and 3–5 styles. Fruit a splitting capsule; seeds shining and compressed. Linen and linseed oil are products of the family.

LINUM—Flax

L. strictum L.   354

A stiff erect annual with small yellow flowers and narrow lance-shaped leaves which are very rough at the edges. Plant 10–60 cm. high; flowers short-stalked, clustered into a flat-topped head or in an elongated spike. Flowers small, ⅓ longer than sepals which are 4–5 mm. long, lance-shaped with long points and twice as long as the pods. HABITAT: dry rocky and grassy places; circum-Medit. May–July.

L. campanulatum L.   84

An attractive flax with large bright yellow flowers with delicate orange veins. A low-growing perennial, 5–25 cm. high with a woody base and spreading angular stems. Leaves with 2 brown glands at their base and with a whitish transparent smooth margin, linear or narrow lance-shaped. Flowers large, 2½–3½ cm. across, with a 'tube' more than twice as long as the calyx. HABITAT: dry hills, limestone; Spain to Italy. April–June.

L. pubescens Russ.   85   PINK FLAX

This pink flax is a common plant of the Lebanon and Palestine. It has large pink flowers, 2½ cm. across, which open wide in the sun and make a fine show of colour. An annual with somewhat bristly hairy stems up to ½ m. high with broad lance-shaped leaves, which are toothed, glandular and with long spreading hairs; the lower leaves are blunt, the upper acute. The flowering branches elongate during flowering; sepals glandular, hairy. HABITAT: fields and stony places; Greece to Palestine, Egypt. March–April.

L. narbonense L.   BLUE FLAX

A perennial flax with a woody base and bluish-grey leaves, and with large blue flowers 3–4 cm. across. Stems stiff, erect, 20–50 cm. high, with the usual narrow lance-shaped leaves of the genus, but the upper floral leaves have thin dry white margins. Flowers on short stalks, shorter than sepals; sepals hairless, lance-shaped with awl-shaped points and white papery margins. Petals egg-shaped with elongated bases, 3 times longer than sepals; stigmas elongated, not club-headed. HABITAT: dry places in the Medit. region; Spain to Italy; Morocco, Algeria. May–June.

L. bienne Mill. (*L. angustifolium Huds.*)   PALE FLAX

This is probably the commonest blue-flowered flax to be found in the

Eastern Mediterranean. An annual, biennial or perennial up to 60 cm. with linear to lance-shaped leaves which are 1-nerved. Petals sky-blue, small, 8–12 mm. long, 2 to 3 times as long as the calyx; sepals egg-shaped, shortly pointed, 1-nerved, sparsely hairy on the inside. HABITAT: meadows, rocky places, glades and hills; circum-Medit. March–June. Pale Flax used to be cultivated in prehistoric times but is now replaced by the cultivated Flax. A British native.

The very similar CULTIVATED FLAX, *L. usitatissimum L.*, is a robuster annual plant with broader, lance-shaped, 3-nerved leaves. Petals sky-blue, about 1½ cm. long; sepals commonly 3-nerved (rarely 1-nerved) with dry papery margins, and with spreading non-glandular hairs at the apex. Widely cultivated in the Mediterranean and not uncommonly naturalized; circum-Medit. March–June. A British casual.

Linen is woven from the stem fibres and linseed oil obtained from the seeds. Flax was probably the earliest plant-fibre used for making clothes: Egyptian mummies were wrapped in linen from the earliest times, and the Greeks and Hebrews used it for shrouds. It was the only important textile fabric in Biblical times. It was widely cultivated in the Middle East but has been largely superseded by cotton today. To obtain linen fibres the fruiting flax plants are cut and steeped in water, or 'retted', for several weeks to separate the fibres from the soft tissues. The fibres are then dried and bleached in the sun.

ZYGOPHYLLACEAE—Caltrop Family

A largely tropical family of herbs and shrubs with usually pinnate leaves and paired persistent stipules. Flowers with parts usually in 5's, and 10 unfused stamens; ovary of 4–5 carpels with 2 or more ovules per cell. The family produces *lignum-vitae*, the hardest and most dense of all woods.

TRIBULUS

T. terrestris L. 355   MALTESE CROSS, SMALL CALTROPS

This is a hairy creeping plant with neat compound leaves and small yellowish flowers. It has an unmistakable fruit which gives the plant its name and recalls the 'caltrops' of classical times, a military weapon which was so devised that however it fell it exposed upwardly projecting spines. The fruits are liable to pierce the feet of animals. Stems 10–50 cm. long; leaves short-stalked with 5–8 pairs of elliptic hairy leaflets. Flowers solitary on short stalks in the axils of leaves, yellow. Fruit 5-starred like a Maltese cross and each part bears 2 long and 2 short hard spines, and several smaller spines. HABITAT: sandy places, fields and track sides; circum-Medit. April–August.

RUTACEAE—Rue Family

Quite an important family which is mainly tropical and which yields valuable essential oils and edible fruits. Mostly trees and shrubs, usually evergreen, with simple or pinnate leaves which are covered with glandular dots and give off aromatic oils. Flowers with 4–5 petals and sepals, and often double the number of stamens; ovary of 4–5 usually fused carpels on a raised receptacle or disk. Fruit a fleshy berry or hard drupe.

RUTA: herbaceous plants with yellow flowers, and petals and sepals 4. Fruit a capsule.

CITRUS: trees or shrubs; flowers white or pink, very sweet scented Leaves simple and fruit large and fleshy.

RUTA—Rue

R. chalepensis L. 80

A yellow-flowered plant with pale glandular foliage which gives off a very strong fetid smell. Stems woody at the base, up to 80 cm. high. Leaves compound, twice cut into oblong elliptic wedge-shaped leaflets. Petals yellow, fringed with long conspicuous upstanding tooth-like hairs. Fruit globular, with 4–5 sharp-pointed lobes close together and not spreading. HABITAT: rocky places, woods, dry banks and thickets; circum-Medit. March–June.

The COMMON RUE, *R. graveolens L.*, is similar but is distinguished by the petals which are smooth-margined or short-toothed. Leaflets oval, strongly aromatic, or narrow-linear and almost odourless. Fruits with rounded lobes. HABITAT: rocks, old walls, dry hills largely on limestone; European Medit., Turkey. May–July. Often cultivated and sometimes naturalized.

A pot-herb with some antiseptic medicinal properties, but its reputation is probably due rather to its strong odour than its medicinal properties. This may be the biblical rue, but this is more likely to be the preceding species, which is a native of Palestine. Used for rheumatism and ear-ache in Palestine; it is also carried as a charm. Shakespeare calls it the 'Herb of Grace'.

CITRUS

C. medica L.   CITRON

A small tree with short stiff spines. Leaves saw-edged, with a leaf-stalk without a wing and not jointed like most other species. Flowers large, 4 cm. across, purplish outside, white within. Fruit ovoid or oblong, very large, 15–25 cm. long, lemon-yellow with a rough fragrant skin and mildly acid flesh. The skin is used for candying.

C. limon Burm. fil.   LEMON

A small tree with stout stiff spines. Leaves small, 5–10 cm. long, with rounded-toothed margins and a leaf-stalk usually with very narrow flanges, and with a prominent joint. Flowers small, 1–2 cm. long, pinkish outside, white within. Fruit oblong to ovoid, 5–10 cm. long, light yellow with a nipple-like swelling and with very sour flesh.

C. maxima (Burm.) Merr. GRAPEFRUIT
A tree with rather glossy leaves, 8–15 cm. long, with shallow indented margins and a broadly winged leaf-stalk. Flowers white, 2–4 cm. long. Fruit very large, 10–15 cm. in diameter, globular, flattened at the ends; skin smooth, pale yellow, flesh acid.

C. aurantium L. SEVILLE ORANGE
A medium-sized tree with oblong-egg-shaped leaves, 5–8 cm. long, pointed and with shallow-indented margins; leaf-stalk broadly winged. Flowers very fragrant. Fruits globular, slightly flattened at each end, 7–8 cm. in diameter, with orange-reddish bitter skin and very sour flesh. Grown for its flowers which are used in perfumery, and for marmalade making; also used as a stock for the sweet orange.

C. sinensis (L.) Osbeck ORANGE
Tree with few or no spines; leaf-stalk narrowly winged. Fruit variable in size, orange to yellow, with sweet flesh and only slightly bitter skin.

C. nobilis Lour. TANGERINE
A small spineless tree with oblong to elliptic wavy-margined leaves, with a narrow winged leaf-stalk. Flowers small, 3 cm. across, white. Fruit ovoid and flattened at each end, 6–7 cm. in diameter, orange-reddish with smooth loose easily peeling skin; flesh very sweet and aromatic.

The citrus fruits, oranges, lemons, tangerines and grapefruits, originate from Asia and are cultivated in many parts of the Mediterranean, particularly on flat coastal valleys and hillsides not far from the sea. They are often planted as ornamental trees, and the scent of 'orange' blossom in the spring is familiar to all who have travelled these shores. The fruits are largely eaten fresh, or crushed to make fruit juices and cordials, or preserved and candied. The flowers are valuable in the scent-making industries, and the flower of the bitter orange forms one of the important ingredients of Eau de Cologne.

MELIACEAE—Mahogany Family

A family of tropical trees and shrubs often with scented wood, and pinnate leaves without stipules. Flowers small, in loose clusters; sepals and petals 4–5; stamens 8–10, fused into a tube surrounding the knobbed style, and often fused to the petals. Ovary superior, 2–5-celled. Fruit a berry or capsule; seeds usually winged. Many useful timbers come from this family such as mahogany, African cedar, rohan tree, toon wood, satin wood, etc.

MELIA—Bead Tree

M. azedarach L. 86, 87 PERSIAN LILAC, INDIAN BEAD TREE
A medium-sized tree with doubly compound leaves which are deciduous, and small lilac flowers in loose clusters amongst the upper leaves. Leaves

large, very variable, 20–50 cm. long, with leaflets egg-shaped to lance-shaped, pointed and toothed. Flowers sweet-scented, about 1 cm. long. Fruit globular, pea-sized, yellow and plump when ripe, but remaining on tree for some time and becoming wrinkled. A native of the Himalaya and China, but commonly planted in the Mediterranean region and naturalized. Flowers in spring.

The seeds of the fruits are used as beads for rosaries, thus giving the name Bead Tree or Arbor Sancta. The leaves, bark and fruit have been used medicinally; fruit somewhat poisonous. The leaves are supposed to discourage mosquitoes.

POLYGALACEAE—Milkwort Family

Usually herbaceous plants, but sometimes shrubs, with simple leaves without stipules. Flowers very asymmetrical with 5 sepals, 2 of which become much enlarged and petal-like; petals 3–5, the 2 outer free, or united with the lower petal; stamens 8, fused into a tube for more than ½ their length, anthers opening by a pore. Fruit a 2-celled pod. In *Polygala* the lower petal is often divided into a brush-like crest of narrow segments, and the stamens occur in 2 sets fused to the petals.

POLYGALA—Milkwort

P. monspeliaca L.
An erect sparsely branching annual bearing narrow lance-shaped pointed leaves and rather elongated terminal spikes of greenish-white flowers. Stem 10–25 cm. high, the upper half flower-bearing. Flowers large, white, 6–7 mm. long; petals much shorter than the 2 large lance-shaped sepals each 6–7 mm. long, with 3 branching veins. Fruit broadly winged, shorter and as broad as the sepals. HABITAT: garigue, rocky places and open woods; circum-Medit. April–June.

P. nicaeensis Risso 367
A perennial with blue, white or pink flowers with the 2 large sepals very much longer than the petals. Stems up to 35 cm. long, flower head soon becoming elongated. Sepals broadly elliptic or obovate and suddenly contracted at the base, 8–10 mm. long, with a network of veins. HABITAT: stony, grassy places; France to Greece and the Islands, Algeria. April–June.

The similar *P. venulosa S. & S.* is readily distinguished by the petals which are ½ as long again as the conspicuous paired sepals. Petals blue, sepals white, netted with green veins; flower heads terminal, few-flowered. Stems woody at the base, frequently branched, with the upper branches rough-hairy. HABITAT: stony places; Greece and the Islands, Cyprus. March–May.

EUPHORBIACEAE—Spurge Family

A large family found throughout the warmer parts of the world. Usually shrubs or herbs with acrid milky juice. Flowers 1-sexed, small, green, with or without petals or sepals; stamens 1 or many. Ovary 3-celled, superior. Rubber, castor oil and tapioca come from this family.

EUPHORBIA: herbs or small shrubs with milky juice and usually spirally arranged, narrow entire leaves. Flowers grouped into heads surrounded by cup-shaped calyx-like bracts (involucre); the heads are stalked – each is called a ray – and they are usually grouped into rounded umbrella-shaped umbels. The so-called flowers are in reality made up of several male flowers, each of one stamen, and a single female flower comprising a 3-celled stalked ovary, and surrounded by green bracts which alternate with glistening yellow glands.

EUPHORBIA—Spurge

E. dendroides L. 90   TREE SPURGE

A robust, woody plant with much branched stems forming large rounded bushes up to 2 m. high. Flower heads terminal, composed of 3–10 rays which are themselves ultimately divided into 2 rays. Stem thick, reddish, naked below and marked with scars of old leaves. Leaves mostly alternate, green, thick, narrow lance-shaped and clustered round the stem. Involucral bracts below flowers not fused, rounded rhomboidal. Glands of flower head yellow, half-moon shaped; fruit smooth, hairless. HABITAT: rocks and hills by the sea; Spain to Palestine, Egypt, Algeria. April–June.

E. spinosa L. 88   SPINY SPURGE

Stems thin, woody and much-branched, forming rounded low-growing rather spiny tufts. Flower heads small, yellow, of 1–5 rays. A spiny hairless shrublet, 10–20 cm. high, with few leaves, each 1 cm. long, narrow lance-shaped and bluish-green. Heads of flowers terminal with each ray ending in a single flower. Glands of flower oval, yellow; fruit about 3–6 mm., globular, hairless and covered with dense conical warty swellings. HABITAT: dry stony, rocky places; European Medit. April–June.

E. acanthothamnos Heldr. & Sart. 89

A low-domed spiny shrub with densely interwoven branches, bright green foliage and yellow umbels of flowers covering the whole bush in the spring. A grey-stemmed much branched shrub, 20–30 cm. high, with old fruiting heads and branches forming sharp spines. Leaves lance-shaped, hairless, soon falling, 1 cm. long. Bracts below flower heads yellow, blunt, obovate; flower heads mostly with 3 rays, each 2–3 times branched. Fruit densely covered with short cylindrical warts. HABITAT: rocky and unculti-vated ground; Greece, Crete and the Islands, Turkey. April.

The domed golden yellow bushes are a familiar sight in spring on the poor open hillsides of Greece.

E. characias L.   93

A tall robust spurge with a terminal mop-head of flowers distinguished at once by its reddish-brown glands and overall hairiness. A perennial plant, 30–80 cm. high, with thick hairy stems arising from a woody base. Leaves leathery, long lance-shaped, narrowed to the base and crowded at the upper part of the stem. Flower head formed of whorls of numerous rays; each ray bears a further pair of rays and rounded, fused bracts. The terminal 'flowers' are encircled by cup-like bracts. Flowers with shallow half-moon-shaped glands with thick blunt rounded tips; fruit covered with many soft hairs. HABITAT: dry places, hills, track sides; Spain to Greece, Crete, Morocco, Libya. March–May.

E. veneta Willd. (E. wulfenii Hoppe)   91

A very robust and upstanding spurge with large terminal domed heads of yellow flowers and stout very leafy stems. A perennial, up to 1½ m. high, with the basal part of the stem woody, naked and scarred with leaf bases, the upper part densely leafy. Leaves narrow lance-shaped, pointed, up to 12 cm. long, and gradually narrowed to the base. The flower head is composed of many lateral heads and a very large terminal head, 15 cm. across, of numerous rays, each bearing rounded bracts from which arise paired terminal rays; bracts surrounding flowers fused into a cup. Flowers with oblong undivided hairy lobes alternating with half-moon-shaped glands. Fruit about 6 mm. long, very hairy. HABITAT: rocky places; Dalmatia, Greece to Rhodes. March–April.

E. biumbellata Poir.

A tall robust spurge with stiff erect stems bearing 2, or sometimes 3, flower clusters placed one above the other and separated from each other by the naked stem. A hairless perennial, 30–80 cm. high, with scattered linear oblong-pointed leaves. Flower heads with 8–15 rays; bracts surrounding flowers more or less rounded; glands half-moon-shaped with 2 long club-shaped horns. Fruit 3–4 mm. long, finely granular. HABITAT: sands and hills by the sea; Spain to Italy, Algeria. April–June.

E. cyparissias L.   CYPRESS SPURGE

Distinguished by its creeping rhizome which puts up many sterile brush-like branches densely covered with narrow, pointed, bright green leaves reminiscent of a young fir tree. A hairless, spreading plant with herbaceous flowering stems up to 50 cm. high, and many shorter sterile stems. Leaves stalkless, 1–3 cm. by 2–3 mm., very numerous. Flower heads of 9–15 rays with kidney-shaped bracts, flowers yellow becoming reddish. Glands of flowers half-moon-shaped with 2 short horns; fruit 3 mm. long, slightly rough, hairless. HABITAT: cultivated ground, hedges and track sides, largely on limestone; European Medit. April–June. Doubtfully native in Britain.

E. myrsinites L.   94

A stiff grey-green perennial with stout stems, and many fleshy broad overlapping leaves, and a rounded head of yellow flowers. Stems spreading at the base, 15–30 cm. high; leaves obovate, broader towards the apex with a short narrow point. Flower heads 7–12 rayed, each bearing 1–2 terminal rays; bracts surrounding flowers kidney-shaped. Flowers with ovate hairy lobes alternating with half-moon-shaped glands with long club-shaped horns often bright reddish-brown in colour. Fruit smooth, hairless; seeds with a rough warty surface. HABITAT: stony places; Corsica, Sicily, Italy, Dalmatia, Greece, Cyprus, Turkey. March–June.

The similar *E. biglandulosa Desf.* (92) has narrow lance-shaped leaves, 3–4 cm. by ½ cm., and rays of flower heads more numerous, and smooth seeds. It is a more robust, more erect and paler grey-silvery plant. HABITAT: stony places; Sicily, Greece to Palestine, N. Africa. February–May.

CORIARIACEAE—Coriaria Family

A family of 1 genus only distinguished by the petals which eventually grow fleshy and enclose the 5 separate carpels. Leaves pinnate-veined.

CORIARIA

C. myrtifolia L. 356
A shrub, up to 3 m. high, with stalkless oval, pointed leaves with greenish flowers in terminal heads, and black shining fruit. Branches greyish, tetragonal in section. Leaves almost stalkless, usually opposite, but some-times 3 or more together, 3-veined. Flowers small, in an erect spike, with 5 conspicuous projecting reddish styles; petals and sepals 5; stamens 10. Fruit of 5 carpels spreading in a star, but with 1 seed. HABITAT: hills and dry woods; Western Medit., Spain to Greece; Morocco, Algeria. April–July.

The berries are poisonous and the foliage is dangerous to stock. The poison may have the same effect as alcoholic intoxication. The bark is rich in tannin and is used for tanning and dyeing.

ANACARDIACEAE—Cashew Family

A mainly tropical family of trees and shrubs with resinous or acrid milky juice and resinous bark. Leaves usually alternate, simple or compound. Flowers small, in clusters, with floral parts in 5's, but stamens usually 10. Ovary 1-celled with 1 ovule, or rarely 2–5-celled; styles 1–3, widely sepa-rated. Fruit usually fleshy with a stone. Cashew nuts, lacquer trees, kaffir dates, hog plums and mangoes come from this family.

PISTACIA: leaves compound of 3–25 leaflets; style on summit of fruit; flowers brownish; petals absent.

COTINUS: leaves simple, untoothed; style placed at the side of the fruit.

RHUS: leaves compound and leaflets toothed; petals and stamens 5, inserted below a disk; flowers greenish or yellowish; plants contain a poisonous milky juice.

118

SCHINUS: leaves with many toothed leaflets; petals 5; stamens 10; plants without milky juice; flowers yellowish or greenish.

PISTACIA

P. terebinthus L. **95, 96**   TURPENTINE TREE, TEREBINTH
A deciduous shrub, 2–5 m. high, with compound leaves recalling an ash tree, but with a strong resinous smell, and globular fruits which are brown when ripe. Leaves of 4–5 pairs of leaflets and a terminal leaflet which are shining above, oval or lance-shaped, and with untoothed margins. Flowers in close compound clusters from end of previous year's shoot, stigmas and anthers reddish-purple. Fruit the size of a pea, red then brown. HABITAT: limestone hills and rocks; circum-Medit. April–July.

From the reddish bark a mild and sweet-scented gum is produced. Large irregular horn-like galls are often found on the plant and they are used for tanning. Often planted over Armenian graves. Theophrastus writes that the best resin is from the Terebinth for it sets firm and is more fragrant.

The similar *P. palaestina* (*Boiss.*) *Post* grows in the Asiatic Mediterranean region and is distinguished by its egg-shaped leaflets which are drawn into a long point, with somewhat hairy margins, and by more spreading and branching flower clusters.

It is this shrub which is probably the turpentine tree and the Teil tree of the Bible.

P. vera L. is the species from which the Pistachio nut is obtained. It is a native of W. Asia and Asia Minor, and is distinguished by its large leaflets, 1–5 in number, which are hairy when young. It was introduced in the Christian era to the Mediterranean and is cultivated in Syria, Palestine and Greece.

P. lentiscus L. **97**   MASTIC TREE, LENTISC.
A dark green spreading evergreen shrub, usually 1–3 m. high, with a strongly acrid, resinous smell, compound leaves and catkin-like flowers. It sometimes grows into a small tree. Leaves of 3–6 pairs of dark green leathery leaflets (terminal leaflet absent), which are broadly lance-shaped, blunt-tipped, shining above, and persist during the winter; leaf-stalks with a narrow green wing. Flowers very small with red anthers in short dense erect spikes in the axils of the leaves. Fruits pea-sized, red then black. HABITAT: a common and widespread plant, abundant in the maquis; circum-Medit. April–June.

A resin is obtained from the punctured stems which is the 'mastic' used in medicine and varnish-making; it is cultivated in the island of Chios for this purpose. Mastic has been used since classical times as a chewing-gum for preserving the gums and sweetening the breath. The Arabs produce an oil from the berries which is edible and used for illumination; it makes a popular sweetmeat called masticha, *and a liqueur known as 'mastiche'. Probably the 'balm' of Genesis and the mastic tree of Susannah.*

119

COTINUS—Smoke-Tree

C. coggygria Scop. (*Rhus cotinus L.*)   357
WIG TREE, SMOKE-TREE, VENETIAN SUMACH

A low tufted deciduous shrub, 1–3 m. high, with rounded aromatic bluish-grey leaves; in fruit it becomes covered in a mass of plume-like reddish hairs giving the appearance of a lawyer's wig. Leaves long-stalked, blunt, broadly orbicular, 4–9 cm. long, dull on both sides. Flowers yellowish, very numerous but mostly sterile, in a widely pyramidal branched head; flower-stalks becoming longer and feathery at maturity. Fruit obovate, 3 mm. long, wrinkled and shining brown when ripe. HABITAT: dry banks, rocks and open woods on calcareous soil; Spain to Syria. May–July.

The wood, known as 'young fustic', is used for dyeing leather. It yields a beautiful orange dye for yarn. The plant smells of turpentine.

RHUS—Sumach

R. coriaria L.   358   SUMACH

A shrub, 1–3 m. high, with compound leaves, hairy branches, milky juice, and whitish flowers in long dense erect spikes. Leaves with 7–15 broadly lance-shaped, thick and toothed leaflets, smooth above and velvety below; leaf-stalks hairy, upper parts winged. Fruit clustered in a spike; berries globular, brown-purple and hairy. HABITAT: dry stony and rocky places, largely on limestone; European Medit. to Palestine, Egypt, Algeria. May–June.

The leaves turn red in the autumn. The juice is poisonous. The bark and the leaves produce a valuable tan used in the preparation of leather, and a yellow dye. The fruits are eaten like capers in the Eastern Medit., and are considered to be both a spice and a medicine.

SCHINUS

S. molle L.   362   CALIFORNIAN PEPPER TREE, PERUVIAN MASTIC TREE

An ornamental tree commonly planted in the Mediterranean region. It is easily distinguished by its neat compound leaves with narrow pointed leaflets, and its red fruits of the size and with the taste of peppercorns, which remain on the trees all the winter.

A large-crowned tree up to 13 m. high with graceful pendulous branches, and pale green leaves with 15–20 pairs of leaflets which are narrowly lance-shaped and toothed. Flowers very small, yellowish-green; fruits hanging in long clusters. HABITAT: it is a native plant of the tropical Pacific coast of S. America. July–August.

The stem produces a gum known as American mastic which is purgative. The leaves have a peppery smell when crushed and are so rich in oil that if fragments of the leaf are placed in water, the oil is expelled with such force that the pieces twist and jerk as if by spontaneous motion.

ACERACEAE—Maple Family

A small family of trees and shrubs with palmate-lobed or pinnate leaves, without stipules and with watery or sugary sap. Flowers small, greenish-yellow, in clusters; sepals and petals 4–5; stamens usually 8, in 2 whorls and attached to the side or edge of a fleshy disk; ovary 2-celled with 2 ovules in each cell. Fruit splitting into 2 winged halves. *Acer saccharinum Wang.* produces maple sugar. Acer is the only genus in the Mediterranean.

ACER—Maple

A. monspessulanum L. 359   MONTPELLIER MAPLE
A small tree with clustered yellowish flowers appearing before, or at the same time as the leaves. Tree or shrub up to 6 m. high with fissured bark. Leaves long-stalked with blade less than 5 cm. long with 3 rounded lobes diverging almost at right angles, hairless on both sides, leathery, shining above and blue-grey below. Flowers yellowish-green in outwardly spreading heads. Fruits hairless, up to 3 cm. long, with wings close together and almost parallel, not spreading. HABITAT: open woods, thickets and rocks, largely on limestone; circum-Medit. April–May.

Theophrastus describes the wood as valuable for making beds and yokes for beasts of burden.

RHAMNACEAE—Buckthorn Family

Small, often spiny, trees and shrubs with simple unlobed leaves and small deciduous stipules. Flowers small, greenish, in axillary clusters. Flowers with cup-shaped receptacle; sepals, petals and stamens usually 5 with stamens opposite petals and arising outside the cup-shaped disk. Ovary attached to base of cup, 2–4-celled with 1–2 styles. Fruit a berry or capsule. Useful products of the family are cascara sagrada, lote fruit, alder buckthorn, coral tree, mabee bark, chew stick and popli-chekke bark.

RHAMNUS: stipules not spiny but branches often ending in a spine; fruit fleshy with 2–3 nuts, small (6–8 mm.), black; leaves pinnately veined; calyx funnel-shaped.

PALIURUS: fruit dry, surrounded by a broad flattened disk; stipules spiny.

ZIZYPHUS: fruit fleshy with 2–3 nuts; stipules spiny and leaves strongly 3-nerved from the base.

RHAMNUS—Buckthorn

R. alaternus L. 363
A shrub with small leathery, evergreen leaves, and heads of small yellowish flowers. A non-spiny hairless shrub, 1–5 m. high, with alternate branches. A very variable plant. Leaves variable in size, dark green, with 4–6 pairs of veins, oval or lance-shaped, with saw-toothed or sometimes untoothed margins. Flowers 1-sexed, very small; sepals acute, erect in female flowers and reflexed in male flowers. Fruit at first red and

then black. HABITAT: rocks, maquis, garigue on limestone; circum-Medit. March–April.

The wood is rank-smelling. Theophrastus writes that the wood is easy to turn and has a whiteness like holly, and the branches are useful for feeding sheep as they are always green.

PALIURUS

P. spina-christi *Miller* **360**   CHRIST'S THORN, JERUSALEM THORN

An extremely spiny shrub with zig-zag branches, oval leaves, and with 2 spines at the base of each leaf, 1 long and straight, the other shorter and downward curving. The fruits are very distinctive and reminiscent of miniature umbrellas or toadstools.

A straggling shrub with long flexible branches and alternate asymmetrically oval leaves, 2–4 cm. by 2–3 cm., shortly stalked, with a 3-nerved blade and shallow-toothed margins. Flowers small, yellow, clustered into axillary bunches. Fruit 2–3 cm. in diameter, umbrella- or disk-shaped with a dry, ribbed and undulating margin. HABITAT: hedges, roadsides and thickets, maquis and garigue; France to Palestine, naturalized in Algeria. June-September.

Possibly Christ's Crown of Thorns was made from this plant.

ZIZYPHUS—Jujube

Z. spina-christi *(L.) Willd.* **361**

Very similar to the Christ's Thorn but with berry-like fruits about the size of a hazel nut and rather dry and astringent to the taste. A shrub or small tree, up to 10 m. high, with smooth white branches bearing pairs of stout unequal recurved spines at the base of each leaf. Leaves alternate, leathery, strongly 3-nerved, 2–4 cm. long, elliptic to egg-shaped with shallow indentations. Flower-stalks woolly; fruit globular, about 1 cm. in diameter. HABITAT: hedges and rocky hills; Asiatic Medit., Egypt, naturalized in Algeria. April–June.

This plant grows commonly around the old city of Jerusalem and is thus more likely to be the plant from which the 'Crown of Thorns' was made than the preceding species. The fruits are edible.

Z. jujuba *Miller,* the COMMON JUJUBE, is a doubtful native in the Mediterranean, but it is sometimes cultivated and occasionally naturalized. Branches green and leaves deciduous; leaves 2½–5 cm. long, egg-shaped, toothed, 3-nerved. Fruit egg-shaped, 2–3 cm. long, purplish, edible, about the size of an olive. Native of China and India.

Z. lotus *L.* is similar with whitish zig-zag branches, smaller (1½ cm. long) broader leaves with almost entire margins, and formidable thorns. The fruits are smaller, drier, globular, about 1 cm. in diameter. Cultivated and extensively naturalized, becoming a pest in Algeria. The fruits are eaten by the poor. HABITAT: dry and stony places, steppes; S. Spain, Sicily, Greece, Cyprus, Asiatic Medit., N. Africa. May–July.

VITACEAE—Vine Family

Mostly climbing shrubs with tendrils which represent modified stems. Leaves simple, palmate or pinnate, with glandular dots frequently present. Flowers in clusters arising opposite the leaf stalks at the nodes. Flowers minute, greenish; sepals, petals and stamens, 4–5, petals often falling early; ovary usually 2-celled with 1 or 2 seeds in each cell. In addition to the extremely important grape-vine there are some ornamental species such as the Virginia creeper and species of Cissus.

VITIS—Vine

V. vinifera L.
COMMON VINE

A woody climber with branched tendrils placed opposite the heart-shaped, deeply indented and lobed leaves. Flowers small, greenish and sweet scented, in dense clusters opposite the leaves. Fruits fleshy, oblong, 5–7 mm. long, generally violet and acid-tasting. HABITAT: commonly cultivated and becoming naturalized near cultivation among rocks, scrub, along river banks and in damp woods; circum-Medit. May–June.

The vine was cultivated by the ancient Egyptians and it is the first plant to be recorded in the Bible as a cultivated plant. The 'fruitful vine' is symbolic of the Jewish people and was commonly cultivated in Palestine in Old and New Testament times. The vine was introduced into Southern Europe largely by the Romans. It is considered to be a native of Armenia and the southern shores of the Caspian Sea. Wine, currants, raisins and sultanas date from antiquity; vine leaves are used for cooking. The lees from wine making are fermented into spirits.

MALVACEAE—Mallow Family

A distinctive family of shrubs and herbs with alternate mostly simple and lobed leaves with palmate veins, small stipules which fall off early, mucilage canals, star-shaped hairs and fibrous stems. Flowers conspicuous, white, yellow, red or purple; sepals and petals usually 5, often with 3 or more extra sepal-like bracts (epicalyx) below the flowers. Stamens numerous and united at the base into a tube often fused to the petals and encircling the styles; stamens with a single 1-celled anther lobe. Ovary commonly 5-celled and splitting up into 1-seeded nutlets. Cotton and many plant fibres are obtained from this family. Many species are ornamental.

MALVA: epicalyx of 3 separate unfused segments arising close beneath the sepals.

LAVATERA: epicalyx of 3 segments fused at their base forming a lobed cup below the sepals.

ALTHAEA: epicalyx of 6–9 segments joined together into a cup.

MALOPE: fruit differing from other genera in having a raspberry-like cone of carpels attached to a rounded axis; epicalyx of 3 broad heart-shaped bracts attached to the flower stalk.

HIBISCUS: epicalyx of 10–12 very narrow segments; fruit oval, of 5 chambers.

MALVA—Mallow

M. sylvestris L. 98 COMMON MALLOW

A rather robust herbaceous plant with broad rounded-lobed leaves and conspicuous rosy-purple flowers irregularly grouped in stalked bunches in the axils of the leaves. A perennial or biennial, 30–120 cm. high, with stalked leaves deeply cut into 5–7 broad lobes which have rounded-toothed margins. Flowers 2½–4 cm. across; petals spreading, rose-purple with darker stripes, broader towards the tip and deeply indented, 2–4 times as long as calyx. Fruit wrinkled, yellowish, more or less hairy, and not enclosed in calyx when ripe. HABITAT: fallow land, track sides, cultivated ground; circum-Medit. April–August. A British native.

M. cretica Cav. has much smaller lilac-blue flowers, 1½ cm. across, and petals shorter than calyx, and long flower stalks. An annual with spreading hairs on the stems; lower leaves rounded with rounded teeth, or 3–5 lobes, upper with 3–5 narrow-toothed segments. Epicalyx 3, narrow lance-shaped like the calyx, densely hairy. HABITAT: stony places, fields and dry hills; Italy to Greece, Crete, Tunisia. April–June.

LAVATERA

L. arborea L. 368 TREE MALLOW

An almost tree-like plant with stout stems, very large rounded, often folded leaves, and large rose-purple flowers veined with deeper purple. Stems woody at the base, up to 3 m. high. A biennial or short-lived perennial. Leaves rounded heart-shaped, hairy, up to 20 cm. in diameter, often folded fanwise, with 5–7 broad, toothed lobes, with fine star-shaped hairs on the upper surface and grey-woolly below. Epicalyx with broad egg-shaped segments fused for ½ their length, longer than the true calyx and becoming greatly enlarged in fruit. Flowers 3–5 cm. across, in simple or compound terminal clusters; petals broad, obovate, overlapping, 2–3 times as long as calyx. HABITAT: rocks and sands not far from the sea; Spain to Greece, N. Africa, naturalized farther east. April–September. Native in Britain.

L. cretica L.

A herbaceous biennial, ½–1½ m. high, with lilac flowers in irregular clusters in the axils of the leaves. Lower leaves rounded heart-shaped, upper cut into 5 triangular, pointed lobes with dense star-shaped hairs on the lower surface. Flowers 3–4 cm. across; petals short-stalked, deeply notched, not overlapping. Epicalyx with long hairs, egg-shaped and shorter than the pointed calyx, not enlarging in fruit. HABITAT: waste places by the

124

sea, track sides, cultivated ground; circum-Medit. March–June. A rare native in W. Britain.

L. maritima Gouan
SEA MALLOW

A woody plant with rounded leaves densely covered with soft whitish hairs, and very pale pink flowers with a crimson-purple centre. Stem up to 1 m. high, with upper leaves less rounded and more angular than the lower. Flowers 3–4 cm. across on stalks as long as or longer than the leaf; epicalyx with narrow lance-shaped lobes, shorter than calyx. Fruit hairless, turning black at maturity. HABITAT: maritime rocks and stony places; W. Medit., Spain to Italy, Morocco, Algeria. February–May.

L. unguiculata S. & S.

A perennial with leaves covered with star-shaped hairs above and densely cottony-hairy below, and large solitary violet flowers. Leaves with a heart-shaped base, deeply cut into 5 oblong, toothed segments with the middle lobe lengthened; upper leaves 3-lobed with a long median lobe and 2 shorter spreading lateral lobes. Petals 2–3 times longer than calyx; epicalyx with rounded tips; fruit rough but not wrinkled. HABITAT: dry places; Greece and the Islands, Cyprus. April–June.

ALTHAEA

A. hirsuta L.
HISPID MALLOW

A branched annual or biennial covered with rough stiff hairs, with solitary long-stalked flowers which are at first rosy-purple and later become bluish. Stems 8–60 cm. high. Lower leaves long-stalked and kidney-shaped, 2–4 cm. across, cut into 5 blunt lobes; upper short-stalked, deeply 3–5-lobed, the lobes becoming narrowest in the uppermost leaves. Flowers large, 2½ cm. across, on long stalks longer than the leaves. Epicalyx with lance-shaped lobes shorter than the calyx which has long pointed narrow lobes. Fruits hairless, rounded on the back and wrinkled; calyx in fruit erect. HABITAT: cultivated ground on limestone; circum-Medit. April–July. A doubtful British native.

A. cretica Weinm.
HOLLYHOCK

A tall wand-like plant with a spike of very large rose or lilac flowers, and large rounded shallowly-lobed leaves. A perennial with a flowering stem 1–2 m. high, covered in dense woolly star-shaped hairs. Leaves stalked, with 5–7 blunt lobes with rounded teeth, with flattened hairs above and dense woolly hairs below. Flowers very large, 6–10 cm. across; epicalyx little shorter than the calyx, both triangular and acute. HABITAT: rocky and grassy places; E. Medit., Yugoslavia to Palestine. May–July.

The very similar *A. pallida* (*W. & K.*) *Nym.*, is distinguished by its narrow bilobed petals conspicuously paler towards the base and by the sparse more or less unbranched flattened hairs on the upper leaf surface. Distribution similar.

The common Hollyhock grown in gardens is *A. rosea* (*L.*) *Cav.* from

China, and is not the same plant that grows wild in the E. Mediterranean; the latter is distinguished by its dense woolly star-shaped hairs and narrower petals. The Hollyhock is frequently cultivated in the Mediterranean region and often escapes and becomes naturalized.

MALOPE

M. malacoides L. 369

A mallow-like plant with handsome deep rose coloured flowers veined with purple. A perennial 10–40 cm. high, covered in long stiff spreading hairs with swollen bases. Leaves oval, blunt, irregularly toothed or lobed, and stalked. Epicalyx of 3 large heart-shaped bracts situated a little below the narrower sepals. Flowers solitary on long flower stalks; petals 2–3 times longer than the lance-shaped sepals. Fruit of many units in a globular head, hairless and wrinkled. HABITAT: waste places; Spain to Turkey, Morocco to Tunisia. June–July. Cultivated medicinally and sometimes naturalized.

HIBISCUS

H. trionum L.   BLADDER HIBISCUS

A mallow-like plant with sulphur yellow flowers blotched with blackish-purple at the base. A hairy annual 15–60 cm. high with lower leaves uncut and upper cut into 3–5 finger-like, lobed segments. Flower 3–4 cm. across; petals twice as long as the calyx which is inflated, papery, bladder-like and hairy. Epicalyx of 12 narrow segments covered with stiff hairs. HABITAT: fields and cultivated ground; native of Eastern Medit. Greece to Palestine, Algeria, naturalized in Western Medit. June–September.

HYPERICACEAE—St John's Wort Family

A family of trees, shrubs and herbs with a resinous yellowish or greenish juice. Leaves opposite and simple and often dotted with pellucid or black glands. Flowers showy, solitary or in loose heads; sepals and petals usually 5; stamens numerous, usually bunched together in bundles; ovary 3–5-celled. Fruit a splitting capsule, rarely a berry.

HYPERICUM—St John's Wort

H. empetrifolium Willd. 370

A low heather-like shrub with narrow leaves in groups of 3's and with typical yellow St John's Wort flowers. Leaves narrow, linear, blunt, with margins rolled inwards and with the upper surface covered with transparent dots. Flowers in an open-branched pyramidal head; petals 3–4 times longer than sepals which are egg-shaped and blunt with black glandular stalkless hairs along the margins. The whole plant smells strongly of resin when crushed. HABITAT: rocky places, in the maquis; Greece and the Islands, Crete, Turkey. May–June.

H. crispum L.

A herbaceous plant, pyramidal in form, with lance-shaped leaves with undulating crisped margins. Stems branched horizontally from the base, 30 cm. high; leaves with heart-shaped bases and blades with transparent glands, and with black spotted margins. Flowers with egg-shaped blunt or pointed sepals without black glands, and petals 4 times as long as sepals. HABITAT: olive groves, vineyards, waste ground; Sicily, Greece to Palestine, Egypt. June–August.

TAMARICACEAE—Tamarisk Family

Small trees or heath-like shrubs with slender branches and small scale- or needle-like leaves. Flowers minute, usually in many-flowered slender spikes, with 4–5 sepals and petals, the latter often becoming dry and withered and remaining on the branches for some time. Stamens many or twice number of the petals, arising from a honey-bearing disk; fruit of 3–4 carpels but 1-celled; seeds with conspicuous and characteristic tufts of hairs. Ornamental shrubs, often planted for shelter on saline soils.

TAMARIX

T. gallica L. 371   TAMARISK

A graceful feathery grey-green shrub with tiny heather-like leaves and long spikes of delicate pink flowers. Much branched, 2–10 m. high with slender reddish branches bearing tiny bluish-grey leaves, without transparent margins. Flowers tiny, less than 2 mm. across, in slender narrow spikes borne on the young branches. HABITAT: on the littoral, damp places, banks of streams; W. Medit., Spain to Turkey, Syria, N. Africa. March–July.

The very similar *T. africana Poir.* is a lower growing shrub (2–3 m. high) with larger flowers (about 3 mm. across) and broader flower spikes borne on last year's branches. Leaves with a papery or cartilaginous margin visible when dry. HABITAT: on the littoral, damp places, banks of watercourses; Spain to Dalmatia, Morocco to Tunisia. March–June.

A related desert species, T. mannifera (Ehrenb.) Bunge, *found in the Sinai desert, produces a sweet gummy exudate when the stems are pierced by a scale-insect. It is probably one of the manna-producing plants of the Bible and the manna is collected by the Bedouins today.*

CISTACEAE—Rockrose Family

Shrubs and herbs with simple, usually opposite leaves with or without stipules, and usually with star-shaped hairs. Flowers showy, solitary or in loose heads. Sepals 3–5; petals 5, often falling during the day of opening; stamens numerous and ovary 1-celled, or with 3–5 partial cells at the base. Fruit a capsule opening by longitudinal splits and with persistent encircling sepals. Many are handsome ornamental shrubs and shrublets.

127

CISTUS: shrubs with large pink or white flowers; fruit on erect stalks and splitting into 5 or 10 valves.

HELIANTHEMUM: herbs or shrublets with small yellow, white or rarely pink flowers; stamens in 2 ranks, all fertile; style elongated, fruit on pendulous stalks, opening by 2–3 valves.

HALIMIUM: closely related to *Helianthemum*; sepals 3 or 5, the 2 outer much narrower than the inner; stigma club-shaped or 3-lobed, stalkless; fruits 3-valved.

TUBERARIA: similar to *Helianthemum*, with yellow flowers, but the style is absent or very short and stamens in 1 rank.

FUMANA: differs from *Helianthemum* by having narrow needle-like leaves and by having a ring of sterile stamens outside the fertile stamens; flowers yellow and stigma distinctly 3-lobed.

CISTUS—Rock Rose

C. villosus L. **100**

A pink-flowered shrub with stalked, egg-shaped or elliptic, rough, often wavy-margined leaves. A densely branched shrub, up to 1 m. high, with leaves 2–3 cm. long, with netted veins on the lower surface and covered on both sides with star-shaped hairs. Flowers pink, hairy-stalked, 4–6 cm. across, 3–5 in an umbel; sepals hairy, 2–3 times shorter than petals, style as long as the stamens. Capsule long-haired, a little shorter than sepals. A very variable species, with or without glandular hairs on the young branches, flower stalks and calyx and the lower surface of the leaves.

C. villosus ssp. creticus (*L.*) *Hay.* (**101**) has the lower surface of the leaves, the young branches, flower stalks and calyx glandular and shaggy-haired. HABITAT: maquis, bushy places, rocks; circum-Medit. March–June.

This plant is probably a source of myrrh, signified by the Hebrew word lôt *of the Old Testament. The word* môr, *also meaning myrrh, refers to* Commiphora myrrha, *a native plant of Arabia and E. Africa.*

The cistus exudes the gum 'ladanum'. It is collected by dragging a kind of rake with leather prongs through the shrubs in the heat of the day. It also adheres copiously to the beards of goats as they browse through the bushes and is collected by the shepherds. Ladanum is a dark brown, fragrant and bitter gum; it is used today in perfumery and medicinal plasters. A kind of tea is made from the leaves.

The similar *C. parviflorus Lam.* has smaller pink flowers, 3-nerved leaves and a very short thick style. Leaves elliptic with sparse hairs above and woolly hairs below. Flowers 3–3½ cm. across, in heads of 2–6. HABITAT: bushy places; Greece, Turkey, Crete, Cyprus. April–May.

C. albidus L. **99**

A conspicuous shrub with whitish-grey velvety leaves and large crumpled rose or magenta flowers. A dense bush, 40–100 cm. high, hardly aromatic, with stalkless paired narrow-elliptic leaves, very hairy on both faces with a network of protruding veins on the lower surface. Flowers 4–6 cm. across in heads of 1–4 at the ends of the branches; petals twice or thrice as long as sepals; stamens orange. HABITAT: garigue, rocky places on limestone soils; W. Medit., Spain to Italy; Morocco, Algeria. April–June.

C. crispus L. 106
A low-growing shrub distinguished by deep pinkish-crimson flowers and pale green wrinkled leaves with undulating margins. A very aromatic plant up to ½ m. high with long-haired stems. Leaves hairy, stalkless, paired and fused together at their base. Flowers 3–4 cm. across, bunched together at the ends of the branches; sepals 5, narrow and long-pointed, little shorter than the petals; style as long as the stamens. HABITAT: garigue, rocks and woods, and often on acid soils; W. Medit., Spain to Italy, Sicily; Morocco to Tunisia. April–June.

C. salviaefolius L. 102
A white-flowered shrub with soft, sage-like, stalked leaves which are not sticky and hardly aromatic. A low spreading bush 30–80 cm. high; leaves elliptic or egg-shaped, green above with star-shaped hairs, white-hairy below. Flowers long-stalked, solitary or in few-flowered heads, buds drooping, flowers 4–5 cm. across, petals white usually with a yellow centre, up to twice as long as the sepals. Outer 3 sepals much broader than the inner 2. HABITAT: woods, thickets, banks, often on acid soils; circum-Medit. March–June.

C. populifolius L. 107
A rather large aromatic shrub with heart-shaped pointed leaves, sticky hairless branches, and large white flowers with a yellow centre. Bushy plant up to 1½ m. high, with hairless stalked green leaves, paler and netted on the lower surface. Flowers in few-flowered heads with drooping crimson buds; flowers 4–6 cm. across, petals 3–4 times as long as the 5 hairy sepals, which become markedly red after petal fall; capsule hairless. HABITAT: thickets, dry places, garigue, rocks; Spain to France, Morocco. May–June.

C. ladaniferus L. 104, 105   GUM CISTUS
A large shrub with very fragrant sticky branches and large solitary white flowers usually with a purple blotch at the base of each petal, sometimes unblotched (notably in S. Spain). Often more than 1 m. high, with stalkless lance-shaped leaves, hairless above and white with cottony hairs beneath. Flowers 5–10 cm. across, shortly stalked and solitary at end of short branches, sometimes giving the appearance of a head of flowers. Sepals 3, hairy, rounded and covered with small swellings. Capsule with 10 chambers. HABITAT: pinewoods, copses and dry hills; W. Medit., Spain to France, Morocco to Algeria. May–June.

The similar *C. laurifolius L.* has broader, stalked, oval lance-shaped leaves and flowers clustered in heads of 3–12. Flowers white with a yellow basal blotch. Sepals acute, hairy; capsule with 5 chambers. HABITAT: thickets, dry places; Spain to Italy, Asia Minor, Morocco. June–July.

C. monspeliensis L. 103

A rather small and very aromatic shrub with small white flowers and dark green narrow lance-shaped leaves rolled under at the edges. Branches with long hairs, sticky; plant usually less than 1 m. high. Leaves stalkless, 3-nerved, wrinkled above, sticky with long hairs on the lower surface. Flowers small, 2–3 cm. across, in erect 1-sided heads of 3–10 flowers; petals white, often with a yellow blotch. Sepals 5, ½ as long as petals. HABITAT: garigue and dry places, often covering large areas; W. Medit., Spain to Greece, Cyprus, Morocco to Tunisia. March–June.

HELIANTHEMUM—Sun Rose

H. salicifolium (L.) Mill.

A small hairy annual with lower leaves oval and with prominent stipules, and an erect lax head of small yellow flowers with the petals scarcely longer than the sepals. Stems 5–20 cm. high; leaves with star-shaped hairs, upper leaves lance-shaped without stipules. Flower stalks about 8 mm. long, spreading horizontally, longer than the sepals and the corresponding bracts. Petals golden yellow without a dark blotch; sepals 5. HABITAT: rocks, grassy places, dry hills; circum-Medit. February–June.

HALIMIUM

H. atriplicifolium (Lam.) Spach 108

A low shrub with whitish scaly elliptic leaves and erect rather lax clusters of large yellow flowers. Flowering stems leafless, spike-like with long-stalked lateral clusters of a few flowers, and covered in dense white hairs and long spreading reddish hairs. Leaves stalked, blade 2–3 cm. long, with a strong central nerve. Flowers about 3 cm. across; sepals lance-shaped with spreading white hairs and a few reddish hairs; capsule hairy at the tip. HABITAT: thickets in the hills, open forests; Spain, Morocco. May–June.

TUBERARIA

T. guttata (L.) Fourr. (Helianthemum guttatum (L.) Mill.) 109

ANNUAL ROCKROSE

A rather delicate hairy annual with lance-shaped, strongly 3-nerved leaves and lax heads of small yellow flowers often with a conspicuous maroon spot at the base of each petal. Stems erect, usually sparsely branched, but very variable, 6–30 cm. high. Basal leaves in rosette but usually dead at flowering time, upper leaves usually with conspicuous narrow stipules. Flower stalks long, slender, bracts often present; flowers 8–12 mm. across; sepals with long hairs and black dots. A very variable plant; some forms are sticky. HABITAT: sunny hills, open thickets, sandy places; circum-Medit. March–June. Native in the Channel Islands and W. British Isles.

T. vulgaris Willk. is a much more showy perennial plant with erect hairless stems bearing a few conspicuous large yellow flowers, arising from a basal rosette of neat 3-nerved hairy leaves. Leaves mostly basal, green above, silvery white beneath, stem leaves few, smaller, hairless, without stipules. Flowers 2–3 cm. across, pale yellow; sepals hairless. HABITAT: pinewoods, garigue and sands; Spain to Italy, Morocco to Tunisia. March–June.

FUMANA

F. thymifolia (L.) Spach ex Webb **372**

A small shrubby rockrose-like plant with narrow, hairy, glandular leaves inrolled at the edges, and small yellow flowers in a lax terminal spike. A perennial 10–20 cm. high, woody at the base; leaves narrow lance-shaped with linear leafy stipules which end in a long hair-point. Flower stalks hairy, a little longer than the sepals. HABITAT: rocks, dry hills, open forests and garigue; circum-Medit. April–June.

CACTACEAE—Cactus Family

A family of plants with green, fleshy, thickened, often jointed stems, almost always without true green leaves but usually with groups of sharp spines. Flowers often large, solitary or clustered, showy; sepals and petals fused into a tube or cup at the base and often attached to the inside; stamens many, also attached to the inside of the tube. Ovary inferior, 1-celled, and usually with many ovules embedded in a fleshy pulp. A family of the desert regions of America; cochineal dye is obtained from beetles living on species of cacti.

OPUNTIA

O. ficus-indica (L.) Mill. **110, 111**   PRICKLY PEAR, BARBARY FIG

This is the most commonly seen 'true' cactus growing in the Mediterranean region. It is a tropical South American plant which was apparently introduced by Christopher Columbus and is now widespread. It often forms impenetrable thickets of spiny swollen stems, 2–5 m. high, of oval flattened racquet-like joints, 10–40 cm. long, placed one above the other and bristling with groups of very sharp spines. Flowers bright yellow, 6–7 cm. across, from the margins of the upper stem joints; sepals and petals numerous, fused into a tube. Fruit ovoid, the size of a hen's egg, brick-red, yellow or purple and edible. HABITAT: dry arid and rocky places; circum-Medit. April–July.

The plant is propagated with the greatest ease and any part of it will quickly root; segments of stems are commonly planted round gardens and habitations, and in a few years it forms dense hedges which keep out animals. The fruits are commonly sold in the markets of the Mediterranean.

131

THYMELAEACEAE—Daphne Family

Trees and shrubs with small simple usually stalkless leaves. Flowers with an elongated receptacle often coloured like the calyx; petals scale-like or absent; parts of flowers in 4–5's; stamens twice this number. Ovary 1-seeded; fruit a nutlet, drupe or berry.

THYMELAEA: flowers small, greenish-yellow; petal-like sepals persisting in fruit; receptacle shortly urn- or cup-shaped.

DAPHNE: flowers larger, conspicuous, sweet-scented; petal-like sepals dropping off in fruit; receptacle cylindrical, often swollen at base; fruit a fleshy drupe.

THYMELAEA

T. hirsuta (*L.*) *Endl.* **112**

A shrublet with white cottony branches and small thick overlapping leaves ranged closely along the stem, and clusters of small yellowish flowers. Much branched, up to 1 m. high, with very tough stems. Leaves tiny, 4–6 mm. by 2 mm., swollen, oval, blunt, concave with white cottony hairs on the lower surface only. Flowers in clusters of 2–5, petal-like sepals blunt, 4–5 mm. long, with silky hairs on the outside; flowers 1-sexed or hermaphrodite. HABITAT: hills, sandy and rocky places not far from the sea; circum-Medit. October–May.

The fibres in the stem are very tough and have been used for primitive rope making.

T. tartonraira (*L.*) *All.* **114**

Quite different in appearance from the preceding species; it has oval, silky, hairy leaves, ranged closely but spreading up the stem. An undershrub up to ½ m. high with the base of the stems covered with swollen leaf scars, and densely leafy above. Flowers small, yellowish, in clusters ranged along the leafy stem. Leaves flat and leathery, 1–2 cm. by 4–7 mm., blunt and narrowed at the base. Flowers 4–5 mm. long, in clusters of 2–5 among the leaves; petal-like sepals with silky hairs on the outside. HABITAT: dry, rocky and sandy places not far from the sea; Spain to Turkey, Cyprus, N. Africa. April–May.

DAPHNE

D. gnidium L. **113**

An erect shrub, 60–120 cm. high, with greyish-white evergreen leaves and white sweet-scented flowers in a terminal leafy head. Branches slender, erect, smooth; young branches covered with soft brown hairs and leafy the whole length. Leaves thick, hairless, narrow lance-shaped, 3–4 cm. by 3–7 mm., with inconspicuous white spots. Flowers in terminal clusters; petal-like sepals white within and darker with silky hairs on the outside. Fruit an ovoid black or red berry. HABITAT: garigue, rocky places and woods; Spain to Greece, Turkey, N. Africa. March–September.

PUNICACEAE—Pomegranate Family

A family with only one genus and two native species in the Orient, but the Pomegranate is now widely cultivated in tropical and sub-tropical regions. The distinctive ovary and the unusual fruit, crowned with a calyx tube, distinguishes it from all other families.

PUNICA

P. granatum L. 364   POMEGRANATE

A small rounded spiny tree with very shiny leaves, large scarlet flowers, and ruddy fruit the size of an apple. Tree, 2–5 m. high, with narrow lance-shaped leathery deciduous leaves, 3–8 cm. by ½–1½ cm. Flowers large, solitary, or up to 3 at the ends of the branches; calyx fleshy, red, persistent; petals 5–8, crumpled and bright scarlet; stamens about 20. Fruit with a hard rind and pink pulpy flesh rather sweetly acid in taste. HABITAT: cultivated and naturalized over the whole Mediterranean region; hedges, rocks and walls. June–July.

The pulp has been used since the days of Solomon for making cooling drinks and sherbets. The rind of unripened fruits gives a red dye used for tanning morocco leather, and the flowers also give a red dye. Pomegranate flowers were used in Persia, Egypt, Rhodes and China as decorative features in art. Supposed to be the fruit Paris gave Venus. From the earliest times it has been a symbol of fertility and there are many references to the pome-granate in the Bible. The bark of the root is valuable for use against tape-worms; the rind of the fruit and the flowers are a powerful astringent.

MYRTACEAE—Myrtle Family

A family of trees or shrubs with undivided smooth evergreen leaves which are dotted with glands and aromatic when crushed. Sepals and petals in 4's or 5's, and stamens numerous in bunches opposite the petals. Ovary inferior forming a fleshy berry or dry capsule in fruit. A family found mostly in the tropics of America and Australia. Useful products are various oils such as eucalyptus oil, timbers such as jarrah from gum trees; guava fruit, allspice, cloves, rose-apple, etc.

MYRTUS: leaves undivided, pinnate-veined; flowers solitary or in few-flowered clusters; petals and sepals 4–5; stamens numerous; fruit 2–3-celled, a berry.

EUCALYPTUS: leaves rigid, undivided, pinnate-veined; flowers 3 or more in umbel or head; receptacle cup-like, fused at base to ovary; petals united with calyx to form a lid (operculum) which falls when flowers open; stamens numerous.

MYRTUS

M. communis L. 374   MYRTLE

An evergreen shrub with small shining dark green box-like leaves, with sweet-scented white flowers and black berries. A close-growing bushy shrub up to 3 m. high, with very aromatic egg-shaped pointed leathery leaves, 2–3 cm. by 1¼ cm. Flowers solitary, 2 cm. across, long-stalked, in axils of leaves; stamens numerous and conspicuous. Fruit a rounded berry turning bluish-black on ripening. HABITAT: maquis, garigue and evergreen thickets; circum-Medit. May–June.

This shrub has been well known since classical times. As a symbol of love and peace it was held sacred to Venus. Wreaths of myrtle were worn by magistrates and by victors in the Olympic games; in Roman times poets and playwrights were garlanded with it. To this day myrtle is worn with orange blossom as a traditional bridal flower, and the Jews still adorn booths with it at the feast of the Tabernacle. It is the 'Myrtle' tree of the Bible. The bark, leaves and flowers produce an oil known as 'Eau d' Anges' used in perfumery; the berries are sometimes fermented to make an acid drink. Myrtle wood is very heavy, it works well, is durable and remarkable for its fine grain; it makes charcoal of the finest quality.

The nymph who turned into a myrtle to escape the attentions of Apollo was called Daphne; this botanically misleading episode results in classically minded Greeks calling the myrtle, Daphne.

EUCALYPTUS—Gums

The large Australian Gum trees are commonly planted in the Mediterranean region on siliceous soils on account of their great beauty and very rapid growth. They can be distinguished by their grey mottled bark, their long hanging blue-grey spear-shaped leaves and fluffy looking flowers which are white, yellow or pink. Of the many cultivated species, the following are the commonest in the Mediterranean region.

E. robusta Smith   SWAMP MAHOGANY
A tall erect tree with a spreading shade-giving crown. Leaves shining green and spreading horizontally. Flowers about 2 cm. across, white, in clusters of 4–12; buds pinkish white with a long pointed operculum.

E. cornuta Labill.
A widely grown species with beautiful shining foliage, and flower buds with long horn-shaped operculum 5–6 times longer than the calyx. The fruits are unusual in having valves prolonged into points more than 1 cm. above the calyx tube and fused together at their tips.

E. viminalis Labill.   MANNA GUM
A tall pyramidal tree with pendulous branches. Foliage bright green, leaves lance-shaped. Flowers 1½ cm. across, 3–8 in a stalked umbel; operculum conical as long as or longer than the calyx. Fruit globose, about 5 mm. across, with a flat rim.

E. globulus Labill. 375   BLUE GUM
A very quick growing, tall tree with greyish-blue foliage. Bark falling off in long strips leaving a smooth greyish-white trunk. Leaves sickle-

134

shaped, balsam-scented and hence used to keep off malaria-carrying mosquitoes. Flowers large, about 4 cm. across, solitary or 2–3 together; bud hard, warty, with a short conical operculum and covered with a bluish-white wax. The most widely planted gum tree; introduced into the Mediterranean about 1860.

E. rostrata Schlecht. **376**   RED GUM

A tall tree with greyish-blue foliage and narrow lance-shaped pendulous leaves. Flowers 5–13 mm. across in compact heads of up to 25 flowers. Operculum prolonged into an acute beak about the length of the calyx.

E. amygdalina Labill.

A tall tree with shining white trunk after shedding its outer bark, and very narrow leaves which, at a distance, give the tree the appearance of a 'head of hair'. Flowers very small; operculum blunt. Planted amongst vines against Phylloxera.

UMBELLIFERAE—Carrot Family

A family of herbaceous plants with often furrowed and pithy or hollow stems, and much-divided leaves. The flowers are usually arranged in flat umbrella-shaped heads (umbels) composed of a number of stalks (rays) arising from the top of the main stem. The rays may end in secondary umbels consisting of small flowers arranged in a similar umbrella-shaped head; the secondary umbels may combine to form a large flat-topped head. The fruits are paired and when ripe separate into 2 1-seeded parts; the ribs and the resin canals on the fruits are important distinguishing features of the different genera. There are many useful plants in this family such as carrot, celery, parsnip, and herbs like coriander, anise, caraway and fennel, while many are poisonous as, for example, hemlock.

ERYNGIUM: flowers in dense rounded heads and surrounded by rigid leaf-like spiny bracts; leaves spiny and thistle-like; not typical of the family. Distinguished from members of the daisy family by the flower structure: the petals are not fused into a tube, etc.

BUPLEURUM: leaves simple and undivided unlike other members of the family; flowers yellow in few-flowered umbels with bracts below the umbels often leafy and conspicuous.

FERULA: very large yellow-flowered plants with large leaves divided many times into narrow linear segments; petals not notched but oval and pointed.

TORDYLIUM: white- or pink-flowered plants with fruits which are hairy or warted and have strongly thickened margins; the outer petals of the outer flowers are much larger than the remainder and they have deeply divided lobes.

ORLAYA: white- or pink-flowered plants with the outer petals of the outer flowers much larger than the rest and very deeply indented with a curved point in the notch; fruit with bristles or spines along the ridges.

SMYRNIUM: erect yellowish-green hairless plants with basal leaves 3 times divided into broad rounded segments; umbels yellow or greenish with few or no bracts; fruit ovoid, flattened.

ERYNGIUM—Eryngo

E. campestre L.

A spiny pale green or bluish-green thistle-like plant with small spiny ovoid heads of purplish or yellowish-green flowers. A perennial, 40–60 cm. high, with diverging branches; basal leaves long-stalked, divided into broadly strap-shaped spiny segments; stem leaves less dissected, clasping, with heart-shaped spiny bases. Flower heads surrounded by 6–7 spreading narrow rigid pale green and sharp-pointed bracts twice as long as the flowers; heads of flowers 1–1½ cm. long; flowers 2–3 mm. across. HABITAT: dry, stony places, track sides, plains and uncultivated ground; European Medit., Turkey, Cyprus, N. Africa. May–August. Rare British native.

E. creticum Lam. (377) is similar but the whole spiny plant is a violet-blue colour. A perennial with many upper divergent branches; lower leaves heart-shaped, soon shrivelling; upper stem-leaves eared at the base and divided into narrow spiny segments. Bracts surrounding flower heads 5, with spines at their base, and 3–4 times longer than the blue globular flower heads. HABITAT: plains, arid regions; Italy to Palestine, Egypt. May–August.

In Greece this plant may become so abundant that it colours large areas of the landscape bluish. The roots are sometimes used against snake bites. The overwintering rosettes are eaten as vegetables by the Arabs. The flowering stems are 'tumble-weeds' and are broken off and blown away by the wind.

BUPLEURUM—Hare's Ear

B. semicompositum L. **378**

This is perhaps the most widespread of the narrow-leaved Hare's Ears in the Mediterranean. It is a branched annual, 5–30 cm. high, with rather dark greyish-green narrow-pointed leaves. Umbels small, 4–6 on long rays of uneven length; bracts below rays and umbels 5, each 3-nerved, narrow, pointed and longer than the umbels. Flowers yellow. Fruits conspicuously stalked, globular, and glandular with little white swellings over the surface. HABITAT: sands and fields by the sea; circum-Medit. April–May.

FERULA—Giant Fennel

F. communis L. **115** (Fig. VIIc)   GIANT FENNEL

A very robust herbaceous plant with large yellow umbels and bright green feathery leaves cut into fine segments. Stem very thick, often branched, hollow, up to 5 m. high but usually 2–3 m. Leaves soft, many times divided into narrow linear segments, the lower with a cylindrical leaf stalk, the upper with large boat-shaped membraneous sheaths. Leaves

136

die down in summer, reappearing in autumn. Central terminal umbel of flowers large with the lateral umbel longer-stalked and over-topping it; bracts below umbels narrow lance-shaped. Fruit elliptic-obovate, 12–18 mm. long. HABITAT: dry hills, walls, waste ground, on limestone; circum-Medit. March–June.

The dried pith is used as tinder; it burns very slowly inside the stem and can thus be carried alight from place to place. Thus Hesiod describes Prometheus as bringing the fire he stole from Heaven 'in a ferula'. In some parts of the Mediterranean the stems are used to make furniture.

The similar *F. chiliantha Rech. fil.* (116) is very spectacular with the upper leaves reduced to large inflated sheaths clasping the stem, and rounded orange-yellow umbels of 25–40 rays (more than in *F. communis*). HABITAT: as for *F. communis*; Rhodes, Turkey.

TORDYLIUM

T. apulum L. 117

A white-flowered umbellifer distinguished by its attractive fruits which are disk-shaped and orbicular and have swollen white borders which are wrinkled or knobbed on the margin. An annual, 20–50 cm. high, with hairy stems, soft below and rough above. Leaves divided into oval, hairy, toothed leaflets with the terminal leaflet as broad as long. Heads of 5–8 umbels with linear bracts, and with the outer flowers of each umbel with a single deeply 2-lobed enlarged petal. HABITAT: cultivated ground, track sides, waste ground; circum-Medit. (not Morocco). April–June.

The very similar *T. officinale L.* (Fig. VIIb) has the outer flowers of each umbel with two large petals deeply divided into two very unequal lobes (*T. apulum* has a single petal divided into two equal lobes). Furrows of fruits with 2–3 resin canals (not one as in *T. apulum*). HABITAT: stony, grassy places; Italy to Greece, Rhodes. April–June.

ORLAYA

O. platycarpos (L.) Koch

An attractive small annual umbellifer with delicate lace-like heads of flowers in which the outer petals of the outermost flowers are very much

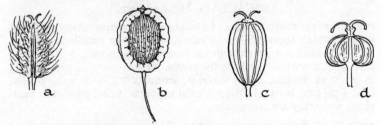

Fig. VII. Fruits of *Umbelliferae*
a *Orlaya grandiflora* (×1½)   **b** *Tordylium officinale* (×1½)   **c** *Ferula communis* (×1)   **d** *Smyrnium rotundifolium* (×2½)

137

enlarged into 2 conspicuous egg-shaped lobes. The fruits are oblong and have 8 ribs running along their length which are set with sharp spines which are reddish in colour. Plant about 30 cm. high. Lower leaves stalked, twice or thrice cut into oblong lobes; the upper leaves have a membraneous sheath and no leaf stalk. Flowers white or pink, the outer petals 2–3 times larger than the inner ones; umbel of 2–3 rays and 2–3 lance-shaped bracts with white papery margins. HABITAT: vineyards, olive groves and uncultivated ground; circum-Medit. April–June.

The similar *O. grandiflora* (*L.*) *Hoffm.* (379, Fig. VIIa) is more showy and has terminal umbels of 5–10 rays, with 5–8 bracts with wide membraneous margins under each head. The outer petal of the outermost flowers is 5–8 times longer than the other petals; fruit shorter and broader, 8 mm. by 5 mm., with whitish spines. HABITAT: calcareous soils and clay in the lower hills; European Medit., N. Africa. April–June.

SMYRNIUM

S. perfoliatum (*L.*) *Mill.*   118

A bright green leafy plant with conspicuous rounded upper leaves clasping the stem, and compound yellow-flowered umbels. A biennial, 30–60 cm. high with furrowed stems and winged ribs; lower leaves with swollen elongated sheath and blades, 2–3 times dissected into oval-oblong segments; upper leaves undivided, oval or rounded with toothed margin. Umbels of 7–12 rays; fruit tiny, 3½ mm. long, black when ripe. HABITAT: olive groves, bushy places; Spain to Palestine, Morocco, Algeria. May–July.

The similar *S. rotundifolium Mill.* (Fig. VIId) has stems without winged ribs and the upper leaves orbicular with untoothed or rarely slightly toothed margins. HABITAT: olive orchards, bushy places; Corsica, Sicily, Italy to Greece, Turkey, N. Africa. April–June.

The similar ALEXANDERS, *S. olusatrum L.*, differs in having the upper leaves divided into 3 segments. HABITAT: bushy places; circum-Medit. April–May. Widely naturalized in Britain.

ERICACEAE—Heather Family

A family of shrublets, shrubs and small trees with simple mostly evergreen leaves, without stipules. Flowers regular, solitary or usually in clusters, with 4–5 sepals and petals, with the latter usually forming a bell-shaped corolla. Stamens mostly double the number of petals, and anthers opening by pores at the tips. Ovary superior, several-celled, with many ovules; fruit a capsule, berry or drupe. Useful products: briar, cranberry, bearberry, and many are ornamental.

ARBUTUS: evergreen trees or shrubs with alternate, stalked, broad leaves; flowers in terminal clusters; corolla 5-lobed and urn-shaped; stamens 10; fruit a spherical berry.

ERICA: evergreen shrubs with small needle-like leaves arranged in whorls; flower parts in 4's; corolla much longer than calyx; stamens 8; fruit a splitting capsule.

ARBUTUS

A. unedo L. **119**   STRAWBERRY TREE

This is a beautiful small tree with laurel-like leaves and clusters of creamy drooping bells. The fruits are unmistakable and reminiscent of strawberries; they are globular, rosy red and rough on the surface. A shrub or small tree up to 10 m. high, with a rough brown bark; the young branches are reddish and rough. Leaves leathery, shining, toothed and lance-shaped. Flowers in branched clusters, creamy white or pinkish, with an urn-shaped corolla and 5 short teeth; calyx teeth triangular. Fruit large, 1½–2 cm. in diameter, with pyramidal swellings over the surface. HABITAT: a characteristic plant of the maquis, rocky places and thickets; Spain to Turkey, N. Africa, Lebanon. October–April. A native plant of Ireland.

In parts of Corsica and Italy the fruits are used to produce an alcoholic distillation; the fruits are edible but not very palatable; the word 'unedo' means 'eat one' and implies that one is enough. Leaves and bark used medicinally. The wood is used for turning and makes good charcoal. In Greece flutes are made from this wood.

A. andrachne L. **120**

A small tree with beautiful smooth red bark, large leathery grey-green leaves, and clusters of whitish flowers. Sometimes a shrub, but usually a tree, up to 5 m. high, with bark peeling in thin sheets leaving a smooth reddish inner bark. Leaves 10–15 cm. long, egg-shaped, with untoothed margins. Flowering shoots upright with glandular hairy branches, flowers white; calyx teeth rounded. Fruit numerous, about 6 mm. in diameter, with a rough netted surface. The leaves of young shoots are hairy and toothed and can be mistaken for *A. unedo*. HABITAT: maquis, woods and thickets; Greece to Palestine. February–April.

Theophrastus writes that women use this wood for parts of the loom.

ERICA—Heather

E. arborea L. **121**   TREE HEATHER

A dense feathery looking shrub with tiny leaves and large terminal pyramidal heads of hundreds of tiny white or pale pink, sweet-scented flowers. Up to 3 m. high with many close erect woody branches; the upper branches are soft and covered with whitish woolly hairs. (In certain places, e.g. Tenerife, it literally grows into a tree up to 7 m. high with stout trunk.) Leaves in whorls of 3 or 4, linear, 3–4 mm. long by less than ½ mm. broad. Flowers with bell-shaped corolla, 3 mm. long, containing the stamens. HABITAT: dominates great areas of the maquis in the W. Medit., largely on siliceous soils; Spain to Turkey, N. Africa; March–May.

The dried branches are used for making brooms, and screens against wind and sun. The wood of the roots makes the best briar pipes; an important honey-producing plant.

E. multiflora L.

A shrub with woody stems and long leaves in close whorls of 3–5 and terminal heads of long-stalked pink flowers with projecting stamens. Stems hairless, pink when young, grey-ribbed in the second year, up to 1 m. high. Leaves hairless, 10 mm. by 1 to 2 mm., with a grey groove beneath. Flowers 2–3 times shorter than the fine flower stalks; corolla bell-shaped, 5 mm. by 2 mm., with anthers and stigmas projecting; calyx 1½ mm. long. HABITAT: dry woods and hills; Spain to Dalmatia, North Africa; July–December.

E. verticillata Forsk. 366

An under-shrub with leaves in whorls of 3 and flowers in pink, compact terminal heads with purple anthers protruding beyond the corolla. A woody perennial, up to 30 cm. high, with hairless branches. Leaves narrow-acute, 5 mm. long and less than 1 mm. broad, grooved beneath. Flowers in clusters of 3–5, forming long spikes towards the ends of the branches; sepals egg-shaped, much shorter (5 times) than the corolla which is 2–3 mm. long and rosy pink; anthers dark purple. HABITAT: on limestone and sandy hills, evergreen thickets and stony places; Dalmatia to Lebanon; August–October.

PRIMULACEAE—Primrose Family

Herbaceous plants largely of the temperate regions. Leaves often whorled and sometimes all basal, simple or lobed. Flowers often solitary or in an umbel. Sepals, petals and stamens 5; petals fused into a tube and stamens fused to it. Ovary 1-celled and many-seeded; fruit a splitting capsule. Many ornamental species.

CYCLAMEN: leaves all basal, heart- or kidney-shaped from a swollen tuber; flowers solitary, nodding; petals reflexed with a short tube at the base.

ANAGALLIS: creeping or erect plants with opposite leaves; flowers borne in axils of leaves, more or less cup-shaped; fruit splitting transversely.

CORIS: leaves linear; flowers in a terminal spike with irregular 2-cleft corolla and papery calyx with 2 rows of teeth.

CYCLAMEN

C. persicum Mill. 126

A spring-flowering cyclamen distinguished from all others by the fruiting stalks which arch downwards instead of curling up in a clock-spring-like spiral. The tuber is spherical and somewhat flattened above and below,

140

with roots arising from the sides and centre of the lower surface. Leaves heart-shaped and more or less pointed, often beautifully marked with grey-green patches and veins, and with fine cartilaginous teeth along the margins. Flowers sweet scented with narrow lance-shaped petals turned upward and commonly slightly spiralled; petals pink, lilac or white with a purple blotch at the throat. HABITAT: rocks, walls and edges of thickets; Greece to Palestine, Crete, Rhodes, Tunisia. January–May.

This is the wild parent of our cultivated winter-flowering cyclamen; the considerable range of variation shown in the wild species has been selected by horticulturists to produce the many forms available to the gardener.

C. repandum S. & S. (C. vernale Mill.) 123

Distinguished by its large usually bright green angular leaves, shallowly cut into wide lobes. It is a spring-flowering species with carmine-red, slightly twisted petals. The tuber is small, 1–3 cm. across, flattened above and below with roots arising only from the centre of the base. Leaves heart-shaped triangular with angular lobes tipped with cartilaginous points. Flowers 1-coloured with no blotch at the throat, 2–3 cm. long, usually pink-purple but rarely white (the form local to Rhodes (123) is commonly white with rose-coloured throat), faintly scented. Fruits on spiral fruit stalks. HABITAT: woods and shady places; France to Greece and Crete, Algeria. March–May.

C. neapolitanum Ten. 124

The flowers appear in summer and autumn before the leaves, and are distinguished by the base of the petals which flare outwards into a pentagonal throat with 10 lobe-like teeth (auricles). Tubers very large, up to 10 cm. across, with roots growing from the upper surface only. Leaves robust and tough, very variable, usually with 5–9 shallow lobes but often rounded or lance-shaped, usually toothed but not cartilaginous; blade green with silvery mottling of great variety; leaf stalks long, creeping and then upstanding. Flowers, 1–1½ cm. long, white to varying shades of rose pink with a dark crimson blotch at the throat; faintly scented. HABITAT: woods and bushy places, largely limestone; France to Turkey, Corsica, Sardinia. July–November.

C. graecum Lam.

An autumn flowering species with beautifully marked leaves with cartilaginous teeth, and flowers with distinct lobes (auricles) at the base of the petals at the throat. The tuber differs from all others by the longitudinal fissures of the corky covering; the roots arise from the base. Leaves very variable, long-stalked, usually triangular-heart-shaped with the margins edged with tiny regular horny teeth, richly and variedly mottled and veined with silvery grey. Flowers deep reddish pink with 2 plum red blotches at the base of each petal; flowers rather large, up to 3 cm. long. Fruit stalk coiled. HABITAT: in bushy stony ground; Greece to Turkey. August–September.

The similar *C. cyprium Ky.* occurs only in Cyprus where it is common. It has auricled flowers which are pure white but with a broad pinkish-purple blotch at the base; flowers medium-sized, up to 2 cm. long, on very short stalks. The margins of the leaf have cartilaginous teeth which are irregular and minutely toothed with a few larger teeth. Autumn flowering and very sweet scented.

C. orbiculatum Mill. (125) is a much smaller-flowered species growing in the lower mountainous regions around the Eastern Mediterranean. It has flowers which are broader than long, and more or less kidney-shaped entire leaves. It flowers in winter and spring.

Some authorities consider that this plant should be renamed *C. vernum Sw.*, others that it should be returned to *C. coum Mill.*

ANAGALLIS—Pimpernel

A. arvensis L. **128**   SCARLET PIMPERNEL, BLUE PIMPERNEL

A well known annual often called the Poor Man's Weatherglass because the flowers open in the sun, and close when it passes behind a cloud. A low spreading hairless plant, 10–30 cm. long, with square-angled stems and opposite stalkless pointed oval leaves. Flowers solitary on stalks in the axil of each leaf; petals longer than sepals, petals scarlet, pink, blue, or rarely lilac. HABITAT: cultivated places, fields and track sides; circum-Medit. March–October. A British native.

There are a number of colour forms and sub-species: *ssp. arvensis Fr.* usually has red flowers, though they may be blue, pink or lilac, and is distinguished by the fruiting stalks which are longer than the leaves; petals overlapping each other with margins densely fringed with 3-celled glandular hairs. *Ssp. foemina* (*Mill.*) *Schinz & Thell.* has fruiting stalks shorter or equalling the leaves; flowers blue, petals narrow, not overlapping each other, with margins with a few 4-celled hairs.

A. monelli L. (127) has large brilliant blue flowers, up to 1½ cm. across, and is a very striking plant. The sub-species *linifolia L.* has linear-pointed opposite leaves, 1 cm. by 1–1½ mm. broad. The flower stalks are many times longer than the leaves. A perennial. HABITAT: bushy places, rocks and grassy places; Spain; Morocco to Libya. March–June. A very variable species.

CORIS

C. monspeliensis L. **122**

A much branched thyme-like plant with compact terminal heads of rose-lilac flowers with reddish-purple sepals, and corolla of 3 large upstanding lobes and 2 smaller lobes. A biennial or perennial, 10–20 cm. high, more or less woody at the base. Leaves many, leathery, linear, blunt, hairless, usually with 2–3 pairs of spines at the base. Calyx bell-shaped, membraneous and inflated, with 2 rows of teeth, the outer 5 curved and spiny, the inner 5 triangular. Corolla 2-lipped; stamens 5; capsule globular.

PLUMBAGINACEAE—Plumbago Family

A family of perennial herbs and shrubs with simple, usually spirally arranged leaves, often forming a basal rosette. Flower heads leafless, often flat-topped and dense; bracts papery. Flowers with parts in 5's with a distinctive, often brightly coloured 'everlasting' tubular calyx which is ribbed, pleated and papery. Corolla often tubular; ovary with 5 styles but with a single ovule.

LIMONIUM

L. sinuatum (L.) Mill. (Statice sinuata L.) **132**   SEA LAVENDER

A handsome perennial with upright winged stems and conspicuous rosettes of wavy-edged leaves, and bright blue-mauve everlasting flowers. A rough plant, 20–50 cm. high, with robust flower stems with generally 3 broad wings which are prolonged at the nodes into lance-shaped leaves; flower head compact, spreading and flattened. Calyx conspicuous, brightly coloured and toothed; the corolla small, whitish and inconspicuous. HABITAT: sandy shores, rocks and track sides by the sea; circum-Medit. March–July.

The flowering heads are cut and dried and will keep their colour for a year or more; widely cultivated for this reason.

STYRACACEAE—Storax Family

Trees or shrubs with entire or toothed alternate leaves, without stipules and often with scaly or star-shaped hairs. Flowers in axillary or terminal clusters; petals and sepals 4–5, stamens fused to base of petals and usually double the number. Ovary superior or inferior, 3–5-celled, with 1 style. Fruit a berry or dry-fleshed drupe. Useful products are storax and gum benzoin.

STYRAX

S. officinalis L. **134**   STORAX

A handsome small deciduous tree or large shrub with pale green foliage, and attractive clusters of rather conspicuous neat white flowers. Much branched, 2–7 m. high, with leaves blunt egg-shaped, bright green, hairless above and with white woolly hairs on the underside, margins untoothed. Flowers 2–4 cm. across, in clusters of 3–6 on stems shorter than the leaves; calyx cup-shaped, with 5 tiny teeth, hairy; corolla fused at base into a short tube with 5–7 lance-shaped petals. Fruit the size of a cherry and covered with a white felt. HABITAT: thickets and woods in the hills; France, Italy to Palestine. April–May.

The gum storax is used to this day in the Roman Catholic church for incense, and for perfumes. It is obtained by making incisions in the branches

143

from which the gum flows. Biblical stacte *and* sweet storax *probably refer to this tree. Rosaries are made from the seeds. The resin known as liquid storax is obtained from* Liquidambar orientalis Mill., *a native tree of the Turkish Mediterranean coast.*

OLEACEAE—Olive Family

Trees or shrubs with usually simple or pinnate leaves without stipules. Flowers in terminal or axillary clusters; calyx 4-lobed; corolla with 4 or more free or fused petals; stamens 2; ovary superior 2-celled. Fruit a drupe, berry, capsule or winged carpel. Useful products are the olive, Manna ash, and timber such as ash and iron wood.

FRAXINUS: trees with simple or pinnate leaves; flowers 1-sexed or hermaphrodite; calyx and corolla 4-lobed or none; fruit 1–2-seeded and prolonged into a broad wing.

PHILLYREA: leaves hairless, simple; corolla tube short, 4-lobed and stamens projecting; fruit fleshy without a hard stone.

OLEA: leaves simple with silvery hairs below; corolla tube short with stamens projecting; fruit fleshy with a hard stone.

JASMINUM: leaves opposite, usually pinnate of 3–7 leaflets; corolla funnel-shaped with stamens shorter than tube; fruit a berry.

FRAXINUS—Ash

F. ornus L. **380** MANNA ASH, FLOWERING ASH

Distinguished by the conspicuous white flowers which are produced at the same time as the young leaves, unlike other species. A tree up to 10 m. high with smooth bark and velvety grey buds. Leaves compound of 5–9 oval or elliptic, stalked and toothed leaflets, green and hairless above, pale or silvery-white beneath. Flowers sweet-scented in large, conspicuous, erect, and later hanging, clusters; petals 4, linear, white; stamens with long filaments. Fruit winged, narrowed and wedge-shaped at the base. HABITAT: hedges, thickets, woods, largely on limestone; European-Medit., Turkey and Lebanon. April–May.

This tree is cultivated in Sicily and Calabria for its 'manna' which is a sweetish exudate containing mannin. This is not the manna of the Exodus which was probably a lichen, but ash manna may well have been sold in the markets during the time of Christ, for this ash is a native tree of Lebanon.

PHILLYREA

P. angustifolia L. **381**

A stiff shrub with opposite pairs of leathery evergreen leaves and small clusters of greenish-yellow flowers in the axils of the leaves. Densely branched, up to 2 m. high with pointed lance-shaped leaves less than 1 cm. broad and without a toothed margin. Flowers fragrant, in rounded clusters,

144

much shorter than the leaves; petals and sepals 4. Fruit a black drupe about the size of a pea, with a short point at the apex. HABITAT: dry hills, rocky stony valleys, abundant in the maquis; largely on limestone; circum-Medit. March–May.

The similar *P. media L.* has oval or elliptic leaves, 8–20 mm. broad, finely toothed and more or less blunt-tipped, and with wedge-shaped bases. A shrub or small tree, up to 4 m. high. Fruit with a short point at the apex. HABITAT: similar to the first.

The similar *P. latifolia L.* has egg-shaped leaves, 6–30 mm. broad, with rounded or heart-shaped bases and with spiny toothed margins. Fruit without a point but slightly indented at the apex. HABITAT: similar to the first.

OLEA—Olive

O. europaea L. 382   OLIVE

The cultivated olive is a tree up to 10 m. high with grey sinuous bark and silvery-grey foliage. Leaves oblong lance-shaped, ashy green above and silky white underneath, and narrowed into a short stalk. Flowers whitish in small erect clusters; calyx cup-shaped with 4 teeth; corolla with 4 spreading lobes. Fruit fleshy, green or black, ellipsoid, with a hard stone.

The sub-species *oleaster* (*Hoffm. & Link.*) *Fiori* is a much-branched shrub with spiny, almost quadrangular branches, with the leaves smaller, 4 cm. long, oblong or egg-shaped, and with small less oily fruits. This is probably the wild form from which the cultivated tree has been derived in antiquity (in Syria). *Ssp. oleaster* grows in maquis, garigue, rocky places, stony hills and walls; circum-Medit. May–June.

Olive oil has been used for cooking, lighting and anointing the body during festivities since the earliest times, and the wealth of many Mediterranean peoples since the Minoans has largely centred round the cultivation of olives. They are frequently mentioned in the Bible and an olive branch was an emblem of peace. Olives preserved and eaten with bread are a staple diet of Mediterranean peasants. Olive oil is also used today for soap making, dressing wool and in medicine, but other cheaper vegetable oils are now taking its place to a large extent and many old orchards are going out of cultivation.

JASMINUM—Jasmine

J. fruticans L. 383   JASMINE

A small erect shrub with shining dark green deciduous leaves and yellow tubular flowers in groups of 1–4 at the ends of the branches. Stems green, usually less than 1 m. high, hairless, ridged and angular. Leaves stalked, single-bladed, or with 3 thick oval and blunt leaflets. Flowers slightly scented, shortly stalked, with a green calyx with awl-shaped teeth, less than $\frac{1}{2}$ the length of the funnel-shaped corolla tube which is $1\frac{1}{2}$ cm. long; fruit a black and shining berry. HABITAT: dry rocky hills, hedges, garigue, largely on limestone; circum-Medit. April–June. Sometimes cultivated.

145

APOCYNACEAE—Dogbane Family

A largely tropical family of herbs, trees and shrubs with milky juice. Leaves opposite or in whorls, blade simple. Flowers clustered or solitary; calyx and corolla with 5 lobes, the latter fused at the base and often with additional segments or lobes; petals twisted in bud. Stamens 5, fused to corolla tube; ovary 2. Fruit of 2 many-seeded carpels, seeds often with a tuft of hairs. There are a few useful tropical members of this family including some used for arrow poisons.

VINCA: flowers large, solitary, with open throat, usually blue; trailing herbs with opposite leaves.

NERIUM: flowers large in terminal clusters, with the throat of the corolla bearing 5 broad-toothed and petal-like structures; flowers usually pink or white; a bush or small tree with whorled leaves.

VINCA—Periwinkle

V. major L. 130   GREATER PERIWINKLE

A trailing perennial with rather thick paired leaves and solitary large blue flowers. Stems trailing, 30–50 cm. long, rooting at the tips. Leaves in opposite pairs, egg-shaped, 5 cm. by 2 cm., rounded and often heart-shaped at the base, blunt tipped and with fine hairs on the margins. Flowers 4–5 cm. across, corolla tubular with 5 spreading, somewhat pointed petals, flower-stalks long but shorter than the leaves; calyx with hairy sepals equalling ultimately the tube of the corolla. Fruit with carpels, 4–5 cm. long, with brown seeds. HABITAT: woods, hedges, shady rocks and banks of streams; European Medit., Turkey; Morocco, Algeria. March–June. Naturalized in Britain.

The similar *V. difformis Pour.* (*V. media L.*) (131) has hairless sepals which are shorter than the corolla tube; acute, pale blue-lilac petals; leaves entirely hairless and rounded but not heart-shaped at the base. HABITAT: hedges, banks and cool places; Spain to Italy, Crete, Morocco, Algeria. February–May.

V. minor L.   LESSER PERIWINKLE

Distinguished by its trailing stems which root at the nodes and by its smaller flowers which are 2½–3 cm. across. Stems 30–60 cm. long, with lance-shaped to elliptic leaves, 2½–4 cm. long, quite hairless. Flowers solitary on stalks longer than the leaves; corolla blue-purple, mauve or white, with blunt petals; calyx teeth hairless, lance-shaped, much shorter than tube of corolla. Fruit of 2 carpels, 1½ cm. long, with blackish seeds. HABITAT: woods, hedges and rocky places; European Medit. Turkey; February–June. Doubtfully native in Britain.

V. herbacea Waldst. & Kit. has herbaceous green non-rooting trailing stems which die back in winter. Leaves hairy with rough margins, narrow-elliptic to linear lance-shaped. Flowers long-stalked, blue-violet, 2½–3 cm. across, petals blunt; calyx with spreading hairs or hairless. HABITAT: thickets and shady places; Yugoslavia to Palestine. February–April.

146

N. oleander L. 129   OLEANDER

A tall robust shrub with narrow grey-green pointed leathery leaves and large sweet smelling pink flowers in terminal few-flowered heads. Branches stiff, erect, up to 4 m. high, with leaves in pairs or in 3's, narrow lance-shaped, with 50–70 pairs of lateral veins. Corolla saucer-shaped with 5 rounded bright pink petals and with the throat of the corolla fringed with long petal-like projections; calyx copiously glandular within at the base. Fruits long cylindrical, ribbed and opening into valves; seeds with a fluffy tuft of hairs. HABITAT: forms thickets along watercourses, wadis, gravelly places and damp ravines; circum-Medit. April–September.

Flowers sometimes white. Commonly cultivated as an ornamental shrub and hedge plant, when the flowers may be red and are sometimes double. The leaves contain a poisonous milky juice. Probably the 'rose growing by the brooks' of Ecclesiasticus. It is known as the 'horse-killer' in India, and is used as a funeral plant in Christian and Hindu regions.

CONVOLVULACEAE—Convolvulus Family

Mostly twining annuals or perennials, but many are erect herbs and shrubs; juice sometimes milky. Leaves alternate and usually simple without stipules. Bracts below flowers often leafy and sometimes protecting buds. Flowers large, showy; sepals 5, petals forming a wide funnel-shaped corolla with 5 stamens attached to it. Fruit a splitting capsule. Useful products are sweet potato, jalap and scammony (the last two are cathartic).

CONVOLVULUS

C. cantabricus L. 137

A non-climbing bindweed with green lance-shaped leaves covered in long, spreading or flattened white hairs, and with pink flowers borne on very long flower stalks. Perennial, 20–50 cm. long, with woody base and much branched ascending stems. Leaves at base lance-shaped with several pronounced nerves on the underside, stem rather silky. Flowering stems longer than the narrow bracts which subtend them, bearing 1–4 stalked pink flowers; corolla 3–4 cm. across, 3 times as long as the hairy calyx. Capsule hairy. HABITAT: dry slopes and rocky places; circum-Medit. April–June.

C. lanuginosus Desr. 138

A plant with narrow silvery leaves and leafless erect stems bearing a rounded head of white, pink-striped flowers. A perennial entirely covered with woolly silvery hairs. Stems not climbing, 10–30 cm. long; leaves linear lance-shaped with a single vein. Flower heads compact, surrounded by lance-shaped leafy bracts longer than the calyx, flowers medium sized, 2–3 cm. across; calyx very hairy, ½ as long as the funnel-shaped corolla. HABITAT: calcareous rocks, rocky slopes; Spain, France, Morocco. March–August.

C. siculus L.

A hairy annual with small blue funnel-shaped flowers, and non-climbing hairy stems and leaves. A spreading or ascending plant, 10–40 cm. long, with egg-shaped leaves slightly heart-shaped at the base and long stalked. Flowers solitary, small, 7–10 mm. across, on a flower stalk shorter than the leaf; corolla blue with a yellow throat. Bracts 2, close under the hairy calyx, and of equal length or longer. Fruit hairless. HABITAT: rocks, screes and dry places; circum-Medit. March–May.

The similar *C. tricolor* L. (133) is distinguished by its oblong leaves gradually narrowed to the base, with the lower stalked and the upper stalkless. Flowers larger, 2–5 cm. across, 3-coloured, yellow, white, and blue from the base upwards, hairy on the veins, borne on stalks longer than the leaves. Bracts some way below the densely hairy calyx. Fruit hairy. HABITAT: hedges, fields, but frequently naturalized near habitation; W. Medit. to Greece; Morocco, Algeria, Tunisia. March–May.

It is a very variable species and forms with larger flowers are widely cultivated.

C. althaeoides L. 136

A purplish pink flowered scrambling perennial with hairy stems and with the upper leaves deeply divided into narrow, very hairy segments. Stems with spreading hairs, climbing up to 1 m.; leaves shortly hairy, green, lower oval with rounded lobes and heart-shaped base, upper cut into 5–9 unequal finger-like segments with the central lobe the largest. Flowers large, 3–4 cm. across, darker in the centre, 1–2 borne on stalks longer than the leaves. HABITAT: bushy places by the sea, hills and cultivation; circum-Medit. April–June.

The similar *C. elegantissimus* Mill. (*C. tenuissimus* S. & S.) (135) has leaves and stem covered in silvery hairs pressed closely to the surface. Leaves as in the previous species, but the upper more silvery and with narrower linear lobes. Flowers solitary, on a stalk much longer than the leaves. HABITAT: dry hills and bushy places; Italy to Turkey, Syria, Morocco, Algeria. April–June.

BORAGINACEAE—Borage Family

A family largely of herbaceous plants, usually with rough hairy foliage. Leaves alternate, undivided, without stipules. Flowers borne on paired outwardly coiled branches with the flowers maturing progressively from the base of the fork. Sepals, petals and stamens 5; petals usually funnel- or bell-shaped with the stamens attached. Ovary deeply 4-lobed with a central style. Fruit usually 4 one-seeded nuts. A family of many ornamental species.

HELIOTROPIUM: flowers white, blue or lilac; corolla lobes separated from each other by a longitudinal fold and often by a small tooth; leaves soft, hairy, stalked or narrowed to the base.

148

CYNOGLOSSUM: corolla funnel-shaped with the throat closed by prominent scales or lobes and hiding the stamens; fruit of 4 flattened nuts covered with hooks or barbed bristles.

BORAGO: corolla with a very short tube and widely spreading petals; stamens projecting forwards in a cone and coming together at the tips.

ANCHUSA: corolla elongated funnel-shaped; throat closed by scales or hairs; nutlets wrinkled; plants rough-hairy, often with swollen bases to hairs.

LYCOPSIS: similar to *Anchusa* but with corolla tube curved near the middle and flowers lop-sided.

ALKANNA: corolla funnel-shaped; throat without hairs or scales but with 5 little transverse swellings alternating with the stamens.

LITHOSPERMUM: corolla tubular or with spreading petals, with hairy longitudinal folds or small scales in the throat; stamens not projecting; nuts very hard.

ONOSMA: plants covered with stiff spiny hairs; corolla cylindrical, straight and widening towards throat with 5 short blunt teeth; throat naked; flower heads leafy.

CERINTHE: more or less hairless plants with somewhat waxy, grey-green leaves; corolla tubular, throat naked; fruit of 2 nuts; flower head leafy.

ECHIUM: corolla funnel-shaped but with an oblique throat and 5 irregular teeth; stamens irregular; plant with stiff rough hairs, stem often speckled.

HELIOTROPIUM—Heliotrope

H. europaeum L. **385**   HELIOTROPE
A hairy annual with arched branches, greyish leaves and white or lilac forget-me-not flowers. Stems 10–50 cm. high, dichotomously branched; leaves shortly stalked, oval, elliptic, green or silvery on both sides and somewhat foetid. Flowers 3–4 mm. across, in tight clusters which are at first spirally coiled and later lengthen; calyx very hairy, divided nearly to the base and spreading in fruit, and remaining on stem after fruit has fallen. Fruit of 4 nuts with small warty swellings. HABITAT: cultivated ground, track sides, waste ground and rocks; circum-Medit. May–September.

CYNOGLOSSUM—Hound's Tongue

C. columnae Ten. **386**
A hairy biennial with erect stems and greyish hairy leaves clasping the stem, and dull purple flowers. Plant about 40 cm. high with a branched flower head and basal leaves lance-shaped, upper leaves blunt, oblong, with heart-shaped base clasping the stem. Flower head free of leaves; flowers 7–10 mm. across. Fruits with an elevated margin and a flat or

concave face furnished with hooked spines and warty swellings. HABITAT: stony ground, waste places; Sicily, Dalmatia to Greece, Turkey. March–July.

The similar *C. creticum Mill.* has pale blue flowers which are veined with violet, and is a grey-hairy plant. The fruits have no elevated margin but convex faces densely covered with short spines mixed with conical swellings. HABITAT: stony ground, thickets and shady places; circum-Medit. March–June.

BORAGO—Borage

B. officinalis L. 139   COMMON BORAGE

A hairy annual with large pure blue flowers opening into a star, and with dark purple projecting anthers. An erect plant, 20–60 cm. high, with thick branched stems. Leaves 10–20 cm. long, egg-shaped, toothed and narrowed into a stalk below, upper leaves clasping stem. Flowers 2 cm. across, long-stalked on leafy branched heads; calyx very hairy, as long as the petals; anthers purple-black, coming together into a conical head. HABITAT: waste places, track sides, gardens, and generally near cultivation; European Medit., Turkey, N. Africa. April–September. British casual.

Cultivated throughout most of the Mediterranean region as a culinary herb. Its medicinal value is doubtful, but it is still used for veterinary purposes in Arab countries.

ANCHUSA—Alkanet

A. hybrida Ten. (A. undulata L.) 143

A hairy biennial with narrow lance-shaped leaves often with undulating wavy-toothed margins and heads of dark blue-purple flowers with velvety tuft in the throat. Stems 30–60 cm. high, leafy and branched at the base with erect dichotomously branching flowering stems with flowers ranged along one side. Basal leaves stalked, upper stalkless, all covered in dense stiff hairs. Flowers small, 7 mm. across, with corolla tube 10–14 mm. long and longer than the calyx; scales in throat of corolla velvety. A very variable species with many local forms. HABITAT: cultivated ground, vineyards and uncultivated places; circum-Medit. March–June.

A. azurea Mill. (A. italica Retz.) 144

A tall herbaceous perennial, 1 m. high, with hairy lance-shaped leaves and loose heads of bright blue flowers. Stem thick, winged, with basal leaves oblong-lanceolate and prolonged into a stalk, upper clasping the stem, all covered in bristly hairs. Flowers 1–1½ cm. across with rounded spreading petals and a brush-like mass of paler blue hairs protruding from the throat. HABITAT: fields, track sides and uncultivated places; circum-Medit. April–August.

The similar *A. strigosa Labill.* has pale bluish violet to almost white flowers, and is a more bristly, rougher plant. The stems and leaves have short, stiff, curved prickly hairs, about 1 mm. long, with broad limpet-shaped bases, scattered over the surface. Flowers 10–15 mm. across, flowering branches much branched. HABITAT: roadsides and fields; Turkey to Palestine. April–June.

LYCOPSIS

Lycopsis variegata L. (Anchusa variegata (L.) Lehm.)   **140, 387**
A rather weak hairy annual with tight clusters of purple or variegated white and purple funnel-shaped flowers and leaves with rough bristly spines. Sparsely branched from the base, up to 30 cm. high; leaves oblong blunt with long bristles on the slightly toothed margins. Flowers small, clustered at the end of long leafy branches, at first pinkish, turning purple and variegated; flowers with a tube longer than the sepals, curved to one side and with irregular rounded petals, sweet scented. HABITAT: cultivated ground, track sides and rough ground; Italy to Greece, Turkey. February–June.

ALKANNA

A. graeca Boiss. & Sprun.
A sticky-glandular leafy plant with bright yellow trumpet-shaped flowers borne in leafy terminal spikes. Stems herbaceous, up to 30 cm. long, covered with glandular and bristly hairs; leaves oblong lance-shaped, the upper half-encircling the stem, covered with bristly flattened hairs. Flower head lax and leafy; calyx bristly and glandular, enlarging in fruit with reflexed lance-shaped teeth; corolla hairless outside, deeper yellow on the tube. HABITAT: rocky places in the hills; Greece. April–June.

The similar *A. orientalis (L.) Boiss.* **(152)** has margins of leaves strongly undulate and crisped. The whole plant is sticky-glandular with a few short bristly hairs. HABITAT: rocky places; Greece to Palestine, Algeria. February–May.

The similar *A. boeotica DC.* has calyx with dense white soft hairs (not glandular-bristly as in *A. graeca*). Stems with long spreading non-glandular hairs and leaves with dense flattened bristly hairs. Corolla golden-yellow or flushed with blue. HABITAT: rocks in the mountains; Greece. April–June.

A. tinctoria (L.) Tsch.   **388**   DYER'S ALKANET
A small bright blue-flowered plant recalling a Lithospermum. A spreading perennial, up to 30 cm. long, with a woody base and a mass of basal leaves. Leaves lance-shaped, the lower stalked, the upper heart-shaped at the base and half-encircling the stem, all leaves covered in rough whitish hairs. Flowers in short spikes on diverging leafy branches, with leafy bracts longer than the flowers; petals about ½ cm. across. Fruit-stalks short and curved downwards after flowering; fruits with irregular rough warts. HABITAT: uncultivated ground, maritime sands, on calcareous soils; circum-Medit. April–June.
The root contains a red dye.

LITHOSPERMUM—Gromwell

L. purpureo-coeruleum L.   147

A rather dark green hairy plant with lance-shaped leaves and striking flowers which are at first reddish-purple and then become a pure dark blue. A perennial with creeping woody stem from which spring long creeping sterile shoots. Flowering stems erect, 25–60 cm. high, with leaves 4–8 cm. by 1–1½ cm., narrowed gradually to the base. Flowers 1½–2 cm. across; flower head at first compact but later elongating. Nuts white and somewhat shining. HABITAT: shady places, hedges and woods, largely on calcareous soils; Spain to Turkey. April–June. A rare British native.

Lithospermum diffusum Lag. (L. prostratum Lois.)   150

A prostrate or scrambling shrub with bright blue flowers. Stems bristly-haired, up to 80 cm. long, with alternate narrow leaves 1½–2 cm. long, 2–4 mm. wide, edges rolled in underneath, rough-hairy on top, paler below. Flowers with 2 cm. long tube opening into 5 rounded lobes, exterior and throat hairy. HABITAT: stony places or among shrubs, often climbing into them: Spain, S. France, Morocco, Algeria. April–July.

ONOSMA—Golden Drop

O. frutescens Lam.   146

An extremely rough, bristly grey-leaved perennial with dense leafy flower spikes of yellow tubular flowers which turn brownish-purple as they mature. Plant woody at the base with many herbaceous stems; leaves oblong lance-shaped, narrowed at the base, with rough bristly hairs (not star-shaped). Corolla hairless, 1½ cm. long, with calyx ⅓ as long, but greatly enlarging in fruit; teeth of corolla shortly triangular and reflexed with anthers protruding. HABITAT: crevices in rocks and cliffs; Greece to Palestine. April–June.

The extremely similar O. echioides L. extends from Spain to Greece and is distinguished by the hairs on leaves which are star-shaped, at least on the lower surface of the upper leaves.

CERINTHE—Honeywort

C. major L. (C. aspera Roth.)   148   HONEYWORT

A striking grey-green, almost hairless annual with oblong, rounded, over-lapping leaves which envelop the stem, and with terminal heads of beautiful tubular yellow flowers. A rather robust plant, 20–40 cm. high, branched above. Lower leaves oboval, narrowed at the base, upper leaves blunt-oval with fine bristly hairs along their margins, all leaves usually sprinkled with white swellings. Flowers large, 2–3 cm. by 1 cm. deeply flask-shaped, hairy, twice as long as the calyx, and with 5 recurved teeth; corolla yellow or with a dark chocolate-purple zone at the base. The species is very variable.

Var. purpurascens Boiss. (149) is a striking form with the broad bracts

152

bearing the flowers a dark purple as well as the flowers. HABITAT: borders of fields and stony places; European Medit., Palestine, N. Africa. March–June.

The similar *C. retorta S. & S.* (151) is also a striking plant with a yellow, violet-tipped corolla and violet bracts surrounding the flowers. The flowers are smaller, 1½ cm. by 3 mm., with slightly incurved points and constricted yellow throats. HABITAT: rocky places; Dalmatia, Greece, Crete. February–June.

ECHIUM—Bugloss

E. italicum L.
A greyish or yellowish biennial covered in dense stiff prickly hairs, and with a broadly pyramidal head of rather small white or flesh-coloured flowers. Stem robust, much branched and bushy, up to 1 m. high. Leaves forming a dense basal rosette, elliptic lance-shaped; stem leaves stalkless with rounded base. Flowers small, 10–14 mm. across, with a regular evenly toothed lip, and with a tube twice the length of the rough hairy calyx. HABITAT: stony places, sands, track sides and derelict cultivations; circum-Medit. March–July.

E. lycopsis L. (*E. plantagineum L.*)   141   PURPLE VIPER'S BUGLOSS
A handsome and striking biennial with bright bluish-violet tubular flowers with projecting stamens, and densely hairy stems and leaves. This very variable plant has stems with erect or spreading branches, up to 60 cm. high; leaves with soft hairs; lower leaves very large, oval, stalked, and soon withering, upper oval-oblong, broader at the base and half encircling the stem. Flowers large, 2–3 cm. long, blue-violet, 3–4 times as long as the calyx, with a throat as broad as long, and with 2 stamens projecting from it. Flowers may be reddish becoming bluish, or white. Fruit 2–3 mm. long, very strongly warted. HABITAT: stony, sandy places, roadsides and waste ground; particularly on siliceous soils; circum-Medit. March–June. A British native.

The similar *E. judaicum Lacaita* (145) is one of the showy plants giving colour to the hills and plains of Lebanon and Palestine during the spring rains. It has larger and more striking flowers produced in greater numbers, often pale blue with pink veins, or violet-purple, the colour depending somewhat on the acidity of the soil; flowers 3–4 cm. long, stamens usually not projecting from the corolla. It is a more bristly plant, and the stalkless stem leaves do not clasp the stem. HABITAT: fields and stony ground; Lebanon and Palestine. March–April.

E. diffusum S. & S.   142
A red or rosy purple flowered bugloss with projecting stamens and with stems and leaves covered in rough spreading hairs. Stems about 30 cm. high, leaves linear to narrow lance-shaped covered usually with stiff hairs and hairs with swollen bases. Flowers in a lax head becoming elongated; corolla hairy, becoming violet when withered, 3 times as long as the calyx

153

with stamens projecting and about as long as the upper lip of corolla. Style bifid, nutlets pointed, with acute swellings. HABITAT: rocky and dry places by the sea; Crete, Greece to Palestine. April–June.

Echium pomponium Boiss.

A very striking summer-flowering bugloss with spikes of flesh-pink flowers on stems 1–1¾ m. tall. Leaves egg- to lance-shaped, 10–20 cm. long, 4–5 cm. broad, forming a rosette, covered with white tubercles. Stem leaves lance-shaped. The 1 cm. flowers are narrowly funnel-shaped, with calyx half as long, and are carried in numerous spikes forming a cylindrical head. The whole plant is covered in white hairs. HABITAT: fields, waste places, especially on clay; Spain, Morocco, Algeria. April–July.

VERBENACEAE—Verbena Family

Herbs, shrubs, trees or woody climbers with usually opposite leaves without stipules. Flowers usually obliquely 2-lipped; calyx usually 5; stamens 4, attached to the fused corolla. Fruit commonly 4-celled but very variable and more commonly a drupe with one or several stones, or a berry. Useful products: teak and fiddle wood.

VERBENA: leaves cut into narrow segments; flower spike narrow, elongated; calyx with 5 teeth; fruit of 4 nutlets.

VITEX: leaves of 5–7 lance-shaped leaflets arranged like the palm of a hand (palmate); stamens projecting beyond corolla; fruit fleshy.

VERBENA

V. officinalis L. VERVAIN

A hairy perennial with square stems, deeply cut leaves, and long narrow spikes of tiny lilac-blue flowers. Stems stiff, erect, up to 60 cm., rough to the touch, woody at the base. Leaves 2–7½ cm. long, deeply cut into oblong egg-shaped lobes, upper leaves narrower and less divided, sometimes entire. Flowers stalkless, at first in a dense spike but elongating in fruit; corolla 4 mm. across, 5-lobed, slightly 2-lipped; calyx 2–3 mm. long, ribbed, hairy. HABITAT: track sides, waste places, screes; circum-Medit. May–September. A British native.

Hippocrates considered this to be one of the few all-curing herbs.

VITEX

V. agnus-castus L. 394 CHASTE TREE

A shrub with hand-shaped (palmate) leaves of narrow pointed leaflets, dark green above and white-felted beneath, and long interrupted spikes of lilac flowers. 1–3 m. high with the young branches quadrangular and white-felted. Leaves deciduous, stalked, with 5–7 lance-shaped leaflets. Flowers small, sweet scented; calyx bell-shaped, white-felted, with 5 triangular teeth; corolla 6–9 mm. long, more or less 2-lipped, hairy outside; stamens projecting well beyond the corolla. Fruit fleshy, 2 mm. across, reddish-black. HABITAT: banks of streams and damp places by the sea; circum-Medit. June–September.

This shrub has, throughout the ages, been associated with chastity. It was well known to the Greeks and the Romans. The seeds from the days of Dioscorides have been celebrated for possessing the power of subduing the 'inclination natural' between the sexes, hence the name Chaste Tree. They were therefore thought to be specially useful to those leading the monastic life and have been called 'Monk's pepper'. However, fresh seeds have an aromatic pungency and have been considered by some to have aphrodisiac properties. It is still used as a medicinal plant for eye diseases and stomach aches, and is the source of a yellow dye. Often cultivated as an ornamental shrub.

LABIATAE—Mint Family

A very distinctive family of herbs and shrubs with quadrangular stems and simple leaves in opposite pairs without stipules, often aromatic and glandular. Flowers arranged in whorls in the axils of opposite, often leaf-like bracts, and the whorls may be bunched together to form an elongated terminal spike. Flowers with calyx 5-toothed, often 2-lipped; corolla tubular often 2-lipped, with the upper lip often large and the lower divided into 3 lobes; stamens 4, attached to the corolla, rarely 2. Fruit of 4 nutlets.

An economically important family from which volatile oils such as peppermint, lavender oil, mint, patchouli, etc., and many culinary herbs such as sage, thyme, rosemary, savory, basil, hyssop, balm and marjoram are obtained.

AJUGA: corolla with a very short upper lip and a conspicuous 3-lobed lower lip; corolla tube with a ring of hairs within.

TEUCRIUM: corolla with no upper lip and lower lip 5-lobed with the middle lobe much larger; tube with no ring of hairs.

ROSMARINUS: stamens 2, projecting; corolla 2-lipped with the upper lip deeply divided and the lower spreading, 3-lobed; leaves narrow with incurved margins.

PRASIUM: stamens 4, projecting; upper lip of corolla convex and arched; calyx 2-lipped; shrublet with oval, toothed leaves.

LAVANDULA: stamens not projecting; flowers blue in a tight spike; corolla tube longer than the calyx.

MARRUBIUM: stamens 4, not projecting; corolla tube shorter than the calyx which has 5–12 small hooked, not spiny teeth; flowers white or yellow; leaves with a network of wrinkles.

SIDERITIS: calyx with 5 spiny teeth; corolla tube shorter than calyx; flowers white or yellow; leaves not wrinkled.

PHLOMIS: flowers large; calyx with 5 equal teeth; corolla with upper lip arched; perennials, often woody, with star-shaped hairs.

155

LAMIUM: calyx soft with 5 equal teeth; flowers rather large; corolla 2-lipped, upper arched, lower 3-lobed and with 2 lateral tooth-like projections; nuts flat-topped.

BALLOTA: calyx funnel-shaped, 10-veined, with 5 long pointed teeth; corolla 2-lipped with upper concave (not arched); nutlets rounded-topped.

SALVIA: calyx 2-lipped, upper with 3 teeth, lower with 2; stamens 2, each with 2 branches, one of which is long and curved, with a single anther; the other branch is short and sterile.

SATUREIA: calyx 2-lipped; corolla 2-lipped with the upper lip flat or convex and lower 3-lipped; stamens diverging at the base but coming together under the upper lip; flowers in a long spike.

MICROMERIA: similar to Satureia but calyx bell-shaped with 13–15 veins and 5 equal teeth; flower in a long spike.

CORIDOTHYMUS: calyx with 13 veins and strongly compressed from above, 2-lipped, the upper lip flat and shortly 3-toothed, the lower 2-cleft; throat hairy; corolla 2-lipped.

THYMUS: calyx 2-lipped, upper lip with 3 teeth, all teeth with hairs; stamens spreading from base to apex; flowers close together in a spike.

AJUGA

A. iva (L.) Schreb.

A very hairy low-growing perennial with a musk-like smell and close clusters of narrow leaves half hiding the purple flowers. Plant with spreading stem, 5–20 cm. long, woody at the base. Leaves greyish-green, covered with whitish hairs, stalkless, narrow lance-shaped, rolled at the edges, without teeth, or with a few teeth towards the tip. Flowers purple, 2–4 in the axil of each leaf, and much shorter than the leaves, but with a tube longer than the calyx. HABITAT: stony, arid places, edges of fields and walls; European Medit., Cyprus, Palestine, N. Africa. March–September.

A. chamaepitys (L.) Schreb. **159**   GROUND-PINE

A low hairy annual, smelling of pine when crushed, with bright yellow 2-lipped flowers set amongst narrow, hairy leaves. Stem branched, spreading from the base and erect at the ends, 5–20 cm. high. Leaves divided to the base into 3 blunt, strap-shaped lobes which lengthen into a leaf-stalk at the base, slightly glandular and sticky. Flowers 9–13 mm. long, paired in the axils of the leaves and much shorter than them; calyx 8 mm. long, bell-shaped, hairy; corolla yellow with a red-spotted lip. HABITAT: fields, dry hills, largely on limestone; European Medit., Turkey, N. Africa. March–August. A British native.

The Arabs use this plant to cure paralysis in animals, and hysteria in horses.

The similar *A. chia (Poir.) Schreb.* has larger, more conspicuous yellow flowers, 20–28 mm. long, streaked with purple, or has entirely purple flowers, 4–5 times as long as the calyx, and projecting beyond the leaves. Leaves 3-lobed, with broadly strap-shaped lobes covered with rough, white hairs. Fruit wrinkled, 4 mm. long. HABITAT: in fields and uncultivated places; Greece to Palestine, Egypt. February–May.

TEUCRIUM—Germander

T. fruticans L.   160, 395

A handsome shrubby germander with small, dark green leaves which are intensely white-felted beneath, and with pale blue flowers. A small shrub, up to 1½ m. high, with entire egg-shaped, stalked leaves, 2–4 cm. by 1 cm., which are tough and remain on the branches during the winter. Flowers solitary, pale blue, 1½ cm. long, long-stalked, arranged regularly round the stem in lax leafy terminal heads. Flowers with stamens and style projecting well beyond the corolla; calyx bell-shaped, white-hairy outside and green within, teeth oval-acute, equal. HABITAT: rocks and woody hills by the littoral; Spain to Dalmatia, N. Africa. February–June.

A very variable species. It is grown as an ornamental hedge plant and can be clipped.

T. divaricatum Sieb.

A shrubby perennial with densely hairy stems and egg-shaped velvet-hairy leaves with long cylindrical, rather one-sided spikes of purple flowers. A branched shrublet up to ½ m. high with rather leathery, stalked leaves with rounded-toothed margins, and paler coloured above. Bracts subtending flowers ovate lance-shaped, shorter than calyx. Flowers in clusters of 2–6; corolla purple, twice the length of the calyx which is bell-shaped, hairy, and with lance-shaped teeth very often purple coloured. HABITAT: rocky places; Greece to Palestine. April–June.

The similar *T. flavum L.* has yellow flowers and less hairy leaves and stem; the calyx teeth blunt, not purple. HABITAT: rocky places; circum-Medit. May–July.

T. chamaedrys L.   396   WALL GERMANDER

A low-growing spreading herbaceous perennial with reddish-purple flowers in a one-sided terminal head. Stem hairy, 10–30 cm. high, with a somewhat woody base, bearing small oval-oblong leaves, 2 cm. by 1 cm., with well-marked rounded teeth, blade rather hard, strongly veined, shining above. Flowers almost stalkless in groups of 2–6 in elongated heads; corolla with lateral lobes acute, calyx reddish, bell-shaped, hairy with more or less equal lance-shaped teeth. HABITAT: dry places, hills and open woods, especially on limestone; circum-Medit. May–June. Introduced into Britain.

Used as a febrifuge.

T. polium L.   389

A white-felted plant with cottony leaves which are rolled inwards at the margins, and with a rounded head of white flowers. A very variable species divided into many micro-species. A perennial, up to 30 cm. high, with woody ascending cottony stems. Leaves blunt-oblong, 1–3 cm. long, stalkless with conspicuous rounded-toothed margins. Flowering heads dense, spherical or ovoid, with flowers longer than the linear bracts. Corolla white or reddish, a little longer than calyx, with stamens hardly projecting; calyx 4 mm. long, bell-shaped, hairy, with egg-shaped teeth. HABITAT: arid places, rocks, screes and sands; circum-Medit. May–August.

A medicinal plant used for stomach troubles, colds and fevers; given in a steam bath.

ROSMARINUS—Rosemary

R. officinalis L. 161   ROSEMARY
Dense evergreen shrub with a characteristic aromatic smell, lavender-like leaves, and lax clusters of lilac-coloured flowers. A much branched densely bushy shrub, up to 1½ m. high, remaining evergreen throughout the year. Leaves 2–3½ cm. by 1–3 mm., leathery, folded inwards along the margin, green and rough above, white-hairy below. Flowers in little clusters towards the ends of the branches. Calyx mealy, 2-lipped, the upper with a single broad oval lobe, the lower with 2 pointed triangular lobes; corolla 2-lipped with the 2 violet stamens and the long curved style projecting from it. HABITAT: maquis and dry hills near the littoral; circum-Medit. In flower throughout the year.

A volatile oil is obtained from this plant and it forms an ingredient for Eau de Cologne, and is used in making hair lotions, cold creams, etc. It is a good honey plant. In Greek and Roman times it was important in religious ceremonies and public festivities. Regarded as a symbol of fidelity. Sometimes used as a conserve and for jam making. It is an antiseptic and an insecticide and used medicinally as an infusion. Often cultivated.

PRASIUM

P. majus L. 398
A small much branched shrub with bright green shining oval, toothed leaves and white or pinkish 2-lipped flowers typical of the family. Stems woody, up to 1 m. high; leaves in opposite pairs, 15–30 mm. long, stalked and heart-shaped at the base. Flowers in pairs in axils of leaves at the ends of the branches; corolla with a short tube and an upper rounded hooded lip, and a 3-lobed lower lip. Calyx conspicuous, bright green and hairless, with 10 veins and 5 sharp-pointed teeth. HABITAT: hedges and rocks; circum-Medit. February–June.

LAVANDULA—Lavender

L. stoechas L. 153   FRENCH LAVENDER
The large purple bracts which stand above the compact head of dark purple flowers make this lavender conspicuous and unmistakable. It is a

158

low, branched, square-stemmed shrub, 30–60 cm. high, with soft narrow grey-green leaves, covered on both sides with dense white velvety hairs. Flower heads in dense spikes, quadrangular in cross-section, and composed of broad, membranous, purple-veined, hairy bracts bearing the flowers; the flower head surmounted by a group of conspicuous sterile enlarged bracts which are brightly coloured. Flowers small, dark purple, 2-lipped; calyx hairy. HABITAT: dry hills, garigue, open woods, on siliceous soils; circum-Medit. February–June.

A very aromatic plant, well known as a medicinal plant in ancient times, and still used in Islamic medicine. It is used in France to perfume linen and to keep away moths.

The similar *L. dentata L.* has narrow leaves with the margins deeply cut into small rounded teeth. Flower spike smaller and crowned with ovate purple bracts. HABITAT: thickets, open woods, rocks on clay and siliceous soils; Spain, Sicily, Malta, Morocco, Algeria. March–June.

L. multifida L. has twice-dissected leaves about 2 cm. long, with narrow pointed hairy segments. Flower spike narrow, compound at base. HABITAT: thickets, stony hills; Spain, Sicily, N. Africa. March–June.

MARRUBIUM—Horehound

M. vulgare L. 390   WHITE HOREHOUND
A greyish-white unpleasant-smelling perennial with paired rounded, toothed leaves, and very compact clusters of small white flowers. Up to 60 cm. high with a stout rhizome and sparsely branched white-woolly stems; leaf-blades 1½–4 cm. long, orbicular, with rounded teeth; wrinkled, green and hairy above, white and hairy below. Calyx with 10 small recurved, hooked teeth; corolla about 1½ cm. long, 2-lipped, with the tube shorter than the calyx. HABITAT: dry slopes, village paths and waste ground; circum-Medit. April–September. A British native.
Used as a stimulant and febrifuge.

SIDERITIS

S. romana L. 391
A small annual with whorls of whitish flowers forming an interrupted leafy spike, and calyx becoming conspicuous in fruit due to the upper much enlarged calyx tooth. A softly hairy leafy plant up to 30 cm. high, with stems branching from the base. Leaves green, oval-oblong with long silky hairs and margins toothed or entire. Flowers in clusters of 6, white or flushed with pink; corolla as long as the calyx, upper lip entire; calyx with 5 teeth, the upper tooth 2–3 times broader than the others.

Ssp. curvidens (Stapf) Holmb. has calyx teeth with long spines which are curved outwards. HABITAT: cultivated ground, track sides; circum-Medit. April–August.
This plant has been used in place of tea.

PHLOMIS

P. herba-venti L.

A perennial with herbaceous stems bearing leathery leaves which are shining above and white-woolly below, and large purple flowers in dense whorls in a leafy interrupted spike. Stems 20–60 cm. high, much branched and downy with spreading hairs, and paired leaves, oval to lance-shaped with rounded-toothed margins. Flower whorls 2–5 and ending in 2 sterile bracts, with 10–12 flowers in each whorl and bracts below whorls stiff, curved and somewhat spiny. Flowers 2 cm. long, purple; corolla hairy; calyx with very sharp spreading teeth. HABITAT: dry places, rocks, track sides and edges of fields, largely on calcareous and clay soils; circum-Medit. May–July.

The similar *P. purpurea L.* (162) has whorls of not more than 8 pinkish-purple flowers and soft spineless hairy bracts; leaves deeply wrinkled on the upper side and white-felted below. It is a dwarf shrub. HABITAT: hills, rocky places; Spain. Spring.

P. fruticosa L. 163   JERUSALEM SAGE

A conspicuous shrub with greyish-white woolly leaves and terminal whorls of large yellow flowers. A shrub up to 1 m. high with white cottony much-branched stems and large oval leaves, 1½–2 times as long as broad, and white and velvety on the underside. Flowering head terminal of 1–3 whorls of 20–30 flowers each; flowers 3 cm. long with a curved hairy 'hood' and a darker yellow spreading 'lip'. Calyx with short rigid recurved teeth; bracts under flower clusters oval-lance-shaped. HABITAT: dry rocky places; France to Palestine. April–June.

The sub-shrubby *P. lychnitis L.* (164) has narrower leaves, 5–6 times longer than broad, and the bracts under the whorls of flowers are much broadened at the base and prolonged to a somewhat spiny point at the apex, and covered with long silky hairs. Flowers yellow, grouped into 4–8 whorls each of 6–10 flowers; calyx teeth 4–5 mm. long, straight, sharp-pointed and silky-haired. HABITAT: rocky places, garigue, open woods, largely on calcareous soil; Spain to France. May–July.

LAMIUM—Dead-nettle

L. moschatum Mill. 154

A rather robust white-flowered dead-nettle with thick hairless stems, and with long calyx teeth spreading into a star in fruit. An almost hairless annual up to 60 cm., with paired, stalked, broadly heart-shaped leaves, blunt-tipped with rounded-toothed margins. Whorls of flowers numerous and many flowered; corolla white, upper lip hairy and tube shorter than the calyx which has triangular lance-shaped teeth enlarging in fruit up to 8 mm. long and spreading outwards. The leaves are sometimes blotched with white. HABITAT: waste places, shady rocks, bushy places and rough ground; Greece to Palestine. January–May.

160

L. maculatum L.   SPOTTED DEAD-NETTLE

A rather strongly smelling dead-nettle with large pinkish flowers and often with a white blotch in the centre of the leaf-blade. A hairy or practically hairless herbaceous perennial, 30–80 cm. high, with leafy stems; leaves stalked, heart-shaped and deeply toothed, the upper with long drawn-out tips. Flowers large, 2–2½ cm. long, with a tube much longer than the calyx. Corolla with upper lip hairy on the outside, with margin fringed with short white hairs and with 2 lateral narrow tooth-like projections; lower lip 2-lobed; anthers purple, pollen bright yellow. A very variable species. HABITAT: open woods, hedges, ditches, waste places by habitations; European Medit., Asia Minor. April–June. Introduced to Britain.

BALLOTA

B. acetabulosa (L.) Benth.

An extremely woolly plant with rounded felt-like leaves and unusual-looking calyces which become spreading, membraneous and umbrella-like in fruit. A perennial with cottony erect stems up to 40 cm. high, and with paired, stalked heart-shaped leaves with rounded-toothed margins, and with wrinkled woolly upper surfaces, and densely white-cottony undersides. The young leaves are quite white, later they become grey. Calyx hairy, opening into a circular, slightly lobed, membraneous funnel; corolla purple. HABITAT: rocky, rough places; Greece and Turkey. Spring.

This plant is sometimes called the False Dittany. The seed vessels are used as a floating wick in olive oil lamps.

SALVIA—Sage

S. triloba L. fil. **155**   THREE-LOBED SAGE

A grey-leaved extremely aromatic undershrub with characteristic sage-like leaves and terminal heads of large violet flowers. Up to 1½ m. high, with a woody base and herbaceous upper branches; the upper leaves often have 2 small lobes at the base thus forming a 3-lobed leaf, but many leaves have only a single oblong-lance-shaped blade. Leaves stalked, 2–5 cm. long, deeply wrinkled above, and covered with short white downy hairs below. Flower spike erect, whorled and sticky, with 4–6 flowers in each whorl; corolla violet, 2 cm. long, and 2 or 3 times as long as the glandular, hairy calyx. HABITAT: bushy and rocky places, maquis; Sicily, Greece to Palestine, naturalized in Algeria. February–July.

An infusion is made from the dried leaves and sold in cafés in the Eastern Medit.; it is called 'faskomelo' in Greek. Apple-like galls develop on the tips of the young branches and they are eaten by the Arabs. An important nectar-giving plant. The green leaves are used to form an infusion with sugar or honey, or eaten fresh. The plant is sometimes burnt to cleanse a habitation.

S. verbenaca L. **156**   WILD CLARY

A sage-like plant with spikes of blue flowers and with large irregularly-lobed or cut leaves mostly from a basal rosette, or from the base of the stem. A perennial with herbaceous stems, up to ½ m. high, with ovate to oblong leaves, 3–10 cm. long, with rounded teeth, wrinkled and sparsely hairy, undivided or shallow-lobed blades. Flowers blue or violet in an erect spike; corolla with unequal lips which diverge a little, the upper lip slightly curved at the end only, darker blue, with style projecting; calyx 2-lipped with angular ribs, glandular and with bristly white hairs. HABITAT: dry hills, screes, track sides, largely on limestone and clay; circum-Medit. February–July. Native to the Channel Islands.

A very variable plant divided into several sub-species, e.g. *ssp. clandestina* (*L.*) *Briq.* with short (10 cm.) nearly leafless stems, and leaves nearly all basal and deeply cut almost to the midrib. Flower heads denser and corolla twice as long as calyx, pale blue with lips very unequal and style protruding. *Ssp. horminoides Pour.* has a robust branched stem, very deeply cut leaves and small bluish or pink flowers, 5–10 mm. long, on a curved spike. Corolla with equal non-diverging lips.

S. horminum L.   392   RED-TOPPED SAGE
A rather small sage with rounded finely toothed leaves and an elongated spike of small pink flowers in whorls, which is usually surmounted by a tuft of brightly coloured bracts. An annual, up to 40 cm. high, simple or branched at the base. Leaves long-stalked, oval, blunt-tipped, 1–2½ cm. broad, finely toothed with rounded teeth, hairy and pale green. Flowers 1–1½ cm. long, pink or violet, 4–6 in a whorl, in the axils of broadly orbicular pointed bracts, with the terminal bracts usually conspicuously coloured, violet or pink. Calyx with glandular hairs, enlarging in fruit. HABITAT: dry places, fields, track sides, always on limestone; circum-Medit. March–June.

SATUREIA—Savory

S. thymbra L.   157   SUMMER SAVORY
A low rough-haired aromatic shrublet with small dark green leaves and pink thyme-like flowers in globular whorls arranged interruptedly on a spike. Stems up to ½ m. high, much branched, rough and hairy; leaves oblong egg-shaped acute, with a wedge-shaped base, blade folded lengthwise and roughly bristled or very hairy. Whorls distinctly globular with many oblong, pointed bracts beneath each cluster; calyx with spreading white hairs and pointed lance-shaped teeth. Corolla pink with the tube shorter than the calyx. HABITAT: dry stony places in the hills; Sardinia, Greece to Palestine. March–May.

A very glandular, aromatic plant. In summer the calyces blacken, looking like 'dark beads strung on thin erect shoots'. A nectar-bearing plant.

MICROMERIA

M. nervosa (*Desf.*) *Benth.*   158
A small perennial shrublet with a woody base and erect stems bearing narrow elongated cylindrical spikes of pink thyme-like flowers. Herbaceous

162

stems rather rough, up to 40 cm. high, hairy on the upper parts. Leaves paired, very shortly stalked, egg-shaped acute, 5–8 mm. long, inrolled at the margins, rough-hairy. Flowers small, 5–8 mm. long, shortly stalked, in many-flowered whorls which separate in maturity into a long interrupted spike; calyx with bristle-like teeth, and long spreading cottony hairs. HABITAT: rocky places in the garigue; Greece to Palestine, Egypt, Algeria. April–July.

M. graeca (L.) Benth. is very similar but it has stalkless leaves, and the upper leaves are much narrower, lance-shaped to linear. Flowers shortly stalked forming lax clusters. Calyx with hairs pressed to the surface, not spreading; throat of calyx hairy. HABITAT: rocky places in the garigue; circum-Medit. April–July.

M. juliana (L.) Benth. has stalkless flowers in very dense whorls forming a long interrupted spike; calyx hairless in the throat. HABITAT: garigue; European and Asiatic Medit., N. Africa. May–July.

CORIDOTHYMUS

Coridothymus capitatus (L.) Rchb. (Thymbra capitata (L.) Griseb., Thymus capitatus (L.) Hoffm. & Link) 393

A low compact cushion-like bushy shrublet with narrow glandular leaves and pink flowers grouped in dense rounded terminal heads with rather broad overlapping bracts which conceal the calyx. Up to 40 cm. high, with young branches with soft hairs, and older stems becoming somewhat hard and spiny. Leaves stiff, narrow-oblong, 2–5 mm. long, glandular-spotted and with hairy margins. Flower clusters globular with egg-shaped or oblong hairy glandular bracts. Flowers commonly pink or purple, rarely white, with stamens projecting a long way beyond the corolla. HABITAT: dry sunny hills; circum-Medit. except France and Corsica. May–September.

It is a source of ethereal oil used in medicine and perfumes. A good honey-yielding plant.

THYMUS—Thyme

T. vulgaris L.

A small, compact grey-leaved bushy shrublet, 20–30 cm. high, with rounded heads of typical pink or whitish thyme-like flowers. A very aromatic plant with densely branched woody stems, and with the young branches velvety white. Leaves narrow lance-shaped, 5–9 mm. by 2–3 mm., covered in glandular dots, and inrolled at the edges, velvety white below. Flowers 4–6 mm. long in a rounded head, but with the lower whorls becoming separated; calyx with bristly hairs. HABITAT: dry slopes, rocks, maquis, always on limestone or clay; Spain to Italy, Morocco. April–July. Cultivated but not naturalized farther east.

This thyme was well known to the ancients and was highly praised by Theophrastus, Horace and Virgil. The leaves produce an oil from which thymol is obtained. The essential oil is used in perfumery and cosmetics and it has many medicinal uses. It is commonly cultivated and used in many culinary dishes as well as for curing hams, drying figs and prunes, etc.

SOLANACEAE—Nightshade Family

Erect or climbing herbs, shrubs or small trees with alternate and variously dissected leaves without stipules. Flowers regular, parts usually in 5's; corolla very variable, fused into a tube but with widely spreading or overlapping lobes; stamens fused to the corolla. Fruit a berry or dry capsule of 2 many-seeded cells.

Very close to the *Scrophulariaceae* but the vascular bundles have food-conducting cells on both sides of the water-conducting cells. There are many useful products from this family: potato, tomato, capsicum, chilli, cayenne pepper and tobacco as well as belladonna and some other poisons and drugs.

HYOSCYAMUS: flowers axillary or in outwardly coiled leafy branches, corolla somewhat asymmetrical; fruit a capsule enclosed in the calyx, and opening by a circular lid.

SOLANUM: petals spreading, flowers not tubular; stamens forming a cone and projecting forward, opening by pores; fruit a berry.

MANDRAGORA: flowers arising direct from a rosette of leaves; corolla bell-shaped with longitudinal pleats; fruit a berry.

DATURA: flower long, funnel-shaped, 5-lobed; calyx long, tubular, often with 5 angles; fruit a spiny splitting capsule.

NICOTIANA: flowers in clusters; corolla tube long and narrow, petals often spreading; stamens at the throat of the corolla but not projecting; fruit a 2-valved capsule.

HYOSCYAMUS—Henbane

H. albus L. **165**   WHITE HENBANE

A coarse pale green hairy sticky plant with large pale greenish-yellow tubular flowers, and a slightly unpleasant odour. A perennial, 20–50 cm. high, with branched erect stems and stalked, egg-shaped, coarsely and bluntly lobed leaves, covered with long soft glandular hairs; upper leaves with heart- or wedge-shaped base. Flowers with a purplish or green throat, $1\frac{1}{2}$–$2\frac{1}{2}$ cm. long; petals rather unequal, and at an oblique angle to the tube; stamens greenish. A variable plant. HABITAT: waste places, track sides, walls and rocks; circum-Medit. March–July.

H. aureus L. **166**   GOLDEN HENBANE

A plant of walls and banks from which it hangs in a mass of yellowish green leaves and spreading stems, bearing large bright yellow flowers with a deep purple throat. A perennial, 30–60 cm. long, with woolly and somewhat sticky stalked leaves, with egg-shaped or orbicular, sharply toothed

164

blades and heart-shaped bases; stem leaves triangular egg-shaped with acute teeth. Flowers 2½ cm. long, with the stamens and style protruding; lobes of corolla unequal, the upper 2 much smaller; flowers all stalked. HABITAT: walls and cliffs; Crete, Rhodes and Turkey to Egypt. March–July.

SOLANUM—Nightshade

S. nigrum L.   BLACK NIGHTSHADE
An annual with egg-shaped leaves and small white starry flowers with petals at first spreading and then reflexed, and yellow protruding anthers. A hairless or hairy plant, 10–60 cm. high; leaves with entire margins or shallowly lobed, wedge-shaped at the base. Flowers 6–8 mm. across. Fruit about 8 mm. in diameter, a black berry or rarely green or greenish-yellow. HABITAT: cultivated ground, waste places, track sides; circum-Medit. January–July. A British native and a cosmopolitan weed of cultivation.

The leaves are boiled and eaten like spinach in Greece. Theophrastus records that the berries can be eaten raw, and in former times the plant was considered worth while growing for this; improved varieties known as 'Garden Huckleberry' are sometimes grown today.

S. sodomaeum L.   **167, 399**
A prickly shrub with very spiny leaves and stems, and with starry violet flowers and yellow shining berries the size of a chestnut. Up to 1 m. high with interwoven branches all covered in stiff sharp yellow spines. Leaves green, becoming hairless above, cut into deep lobes with the veins strongly spined on both surfaces. Flowers large, calyx hairy and spiny; corolla hairy, 3 times as long as calyx. Fruit a berry about 2 cm. across, shining yellow. HABITAT: pastures, sands, track sides and rubbish tips; Western Medit., Spain to Greece, Morocco to Libya. May–August. Sometimes cultivated and naturalized in Greece and E. Mediterranean.

S. incanum L. (S. sanctum L., S. coagulans Forsk., S. hierochunticum Dun.), the PALESTINE NIGHTSHADE or JERICHO POTATO, is very similar but has leaves which are wavy, not cut into lobes, and are densely grey-woolly. HABITAT: Old walls, roadsides, waste places; Palestine. May–July.

An error of naming has presumably crept in here, for with its habitat *S. incanum* should presumably be the 'Apple of Sodom', not *S. sodomaeum.*

MANDRAGORA—Mandrake

M. officinarum L.   **169, 170**   MANDRAKE, LOVE APPLE
This famous plant is said to give out an unearthly shriek when pulled up, and there are many classical and historical references to it. From a thick forked taproot large wrinkled egg-shaped leaves spread over the ground in a large irregular rosette with violet flowers or orange fruits in the centre. Leaves 15–25 cm. long at flowering time, but subsequently enlarging up to 40 cm. long, lance- or egg-shaped, dark green, minutely hairy, often with crisped, toothed, wavy margins. Flowers short-stalked, bell-shaped with

5 petals. Fruit a globular yellow or orange fleshy berry, 2½–3 cm. in diameter. There are spring- and autumn-flowering forms: the former have elliptic fruits, the latter globose fruits. HABITAT: deserted fields and stony places; Spain to Palestine (not France), N. Africa. Spring and autumn.

The fruit has a peculiar smell and is considered by some to be a delicacy. The Mandrake of the Bible. The plant is slightly poisonous and was much employed in medicines in ancient times. Has long been famous in love-potions and said to have some amorous and aphrodisiac properties. In the Middle Ages it was a popular pain-killer during operations.

DATURA

D. stramonium L.   THORN-APPLE
A stout branched leafy plant with coarsely toothed egg-shaped pointed leaves and solitary large white trumpet-shaped flowers, and spiny green fruits resembling horse-chestnut fruits. A hairless annual, 40 cm. to 1 m. high; leaves up to 20 cm. long. Flowers 6–10 cm. long, erect; calyx about 4 cm. long, pale green and 5-angled with 5 narrow triangular teeth; corolla usually white, sometimes purple, lobes narrow, long-pointed. Fruit 4–5 cm. long, ovoid erect. HABITAT: place of origin unknown but now cosmopolitan; cultivated places, waste ground; circum-Medit. April–September. A British casual.

A very poisonous and narcotic plant which contains the alkaloids hyoscyamine, hyoscine, and scopolamine. Theophrastus writes – 'If $\frac{3}{20}$ of an ounce is given, the patient becomes sportive and thinks himself a fine fellow; twice this dose, the patient goes mad and has delusions; thrice this, for permanent insanity, and four times the dose he is killed!'

D. metel L. (168) has larger white or slightly rose-coloured flowers, 15–20 cm. long. A greyish densely hairy plant with oval leaves with uncut or shallowly indented margins. Tube of corolla purple; calyx hairy; fruit spiny, pendulous. HABITAT: a plant of S. America, cultivated and naturalized in parts of the Mediterranean on waste ground, cultivated ground and gravel. Flowers spring and summer.

NICOTIANA—Tobacco

N. glauca Graham   400
A woody plant from South America with long tubular yellow flowers in a loose head, and egg-shaped to elliptic, pointed blue-grey leaves. A shrub, 2–3 m. high, with erect sparsely branched, hairless blue-grey stems. Leaves tough, hairless, on a long stalk longer than the blade. Flowers with a corolla tube 3–4 cm. long and about 5 mm. in diameter, hairy on the outside and with 5 short teeth. Calyx hairless, tubular with 5 pointed teeth and more or less enclosing the fruit. HABITAT: thoroughly naturalized on old walls, rocks and waste ground throughout the Mediterranean region. In flower nearly all the year round.

166

SCROPHULARIACEAE—Figwort Family

Mostly herbs and shrubs with leaves without stipules. Flowers largely symmetrical in one plane only with usually 5 sepals, petals and stamens. Corolla very variable, bell-shaped or tubular, with spreading lobes or strongly 2-lipped. Stamens attached to corolla, often 4, sometimes 2. Ovary superior, 2-celled, each with many ovules; fruit a splitting capsule. A family of many ornamental plants.

VERBASCUM: flowers clustered into long terminal spikes; corolla a spreading 5-lobed cup; stamens 5.

ANTIRRHINUM: corolla strongly 2-lipped with the lower lip 3-lobed and a boss-like 'palate' closing the throat; capsule opening by 3 pores.

LINARIA: corolla strongly 2-lipped with a spur projecting downwards; capsule opening by 4–10 apical valves.

SCROPHULARIA: corolla dull-coloured, a nearly globular tube with 5 small lobes; fertile stamens 4, with usually 1 sterile lobe; plant often square-stemmed.

VERONICA: flowers blue, pink or white; corolla with a very short tube and 4 spreading rounded petals; stamens 2.

BELLARDIA: flowers 2-lipped with the upper lip forming a hood and the lower 3-lobed; calyx swollen bell-shaped, with 4 blunt lobes 3 or 4 times shorter than rest of calyx; seeds with strong longitudinal ribs or wings.

PARENTUCELLIA: similar to *Bellardia*: calyx cylindrical with sharp teeth.

VERBASCUM—Mullein

V. sinuatum L.

A whitish or yellowish green woolly biennial plant with overwintering rosettes of short-stalked, somewhat lobed, undulating-margined leaves. Flower stem robust, up to 2 m. high, with slender branches bearing interrupted clusters of rather small yellow flowers 1½–2 cm. across in short-stalked groups of 2–5; stamens with violet-purple hairs. HABITAT: bushy places, pasturage; circum-Medit. May–August.

The similar *V. undulatum Lam.* **(176, 177)** has large overwintering rosettes of leaves covered in golden-yellow hairs and with wavy incurled edges. Stem leaves egg-shaped with a wide enveloping base and almost entire margin. Flowers bright yellow, 2–5 cm. across, in distinct clusters; stamens with white or yellow hairs. HABITAT: rough dry ground, rocky places; Dalmatia to Greece and the Islands; Turkey. Spring.

Mulleins of this type are largely left by grazing animals and have probably become much more common since mankind has cleared and depauperated the land. They are used as fish poisons in Greece.

V. spinosum L. is a low, much-branched dense shrub with spiny branches. Leaves thinly hairy, oblong and prolonged into a stalk. Flowers solitary,

167

stalked, small, yellow; anthers with yellow hairs. HABITAT: rocky, bushy places by the sea; Crete.

ANTIRRHINUM—Snapdragon

A. majus L. SNAPDRAGON

This large-flowered snapdragon, which is so well known in cultivation, is a native of the Mediterranean region. A short-lived perennial, up to 80 cm. high, with simple or branched stems bearing many lance-shaped hairless leaves about 5 cm. long. Flowers borne in dense terminal spikes, corolla large and showy, reddish-purple touched with yellow on the throat-boss, 3–4 cm. long, and several times as long as the calyx which has egg-shaped acute lobes. Fruit longer than the calyx; fruit stalks and calyx glandular, hairy. HABITAT: rocks, old walls and dry places; circum-Medit. March–July. Naturalized in Britain.

The closely related *A. latifolium Mill.* (171) has broader oval, blunt, hairy leaves, and stem and leaves glandular-hairy. Flowers large, yellow, rarely purple with a swollen boss at the base of the corolla tube. HABITAT: rocks and dry places; Spain to Italy. April–July.

A. orontium L. (*Misopates orontium (L.) Raf.*) **172** WEASEL'S SNOUT

A hairy annual with small pink snapdragon flowers. Erect and usually simply branched plant, 20–50 cm. high, with hairy, glandular, linear or narrow elliptic-pointed leaves, 3–5 cm. long. Flowers solitary, more or less stalkless, in the axils of long leafy bracts; corolla pinkish-purple, 10–15 mm. long, with a yellowish throat-boss. Calyx lobes linear, unequal in length, and generally longer than the corolla. HABITAT: cultivated ground, track sides, sandy places, walls, vineyards; circum-Medit. March–June. A British native.

LINARIA—Toadflax

L. triphylla (L.) Mill. **173**

A robust hairless annual with erect stems bearing leaves in groups of 3, and medium-sized, 3-coloured flowers. Stem thick, up to 40 cm. high, branched only at the base; leaves whorled, 3-veined, broadly egg-shaped. Flowers borne in a cylindrical, rather close spike which elongates in fruit; corolla and spur 1½–2 cm. long, pale yellow with orange throat boss, spur 8–10 mm. long, flushed with purple; calyx hairless with oval lobes about equalling the fruit. HABITAT: fields on the littoral, cultivated ground, vineyards; Spain to Greece and the Islands, Palestine, Morocco to Libya. February–June.

L. hirta Moench (174) is another robust annual species, hairy and glandular above, more or less hairless towards the base. Stem thick, up to 45 cm. high, sometimes branched near summit, with rather bluish elliptical leaves, opposite at base of plant, alternate higher up. Flowers borne in a close, cylindrical spike; corolla and spur 2–2½ cm. long, white with

168

orange blotch on throat-boss and faint lines on upper segments, spur 12–15 mm. long, whitish. HABITAT: fields, waste places; S. Spain. April–June.

L. chalepensis (L.) Mill. 401

A delicate white or very pale yellow-flowered annual with small flowers, long thin curved spurs and narrow strap-shaped leaves. An unbranched or sparsely branched plant, 15–40 cm. high, with thin stems and linear pointed leaves, 4 cm. by 2 mm., and 1-veined; lower stem leaves and those of the short sterile shoots in 3's or opposite. Flowering spike long and lax with flower stalks shorter than calyx. Flower, including spur, 10–12 mm. long, spur long thin and gracefully curved; calyx with linear segments elongating to over 1 cm. and much beyond the fruit. HABITAT: sandy fields and rocky places; Spain to Palestine, Egypt. April–July.

L. pelisseriana (L.) Mill. 402

A very delicate plant with violet flowers with a whitish throat-boss and a long straight awl-shaped spur. An erect unbranched annual, up to 40 cm. high, with stem leaves very narrow, strap-shaped and pointed, and leaves on sterile shoots oval. Flowers at first in a close head which later elongates; corolla including spur 12–18 mm. long, spur about 6 mm. long, intense violet in colour. Calyx with acute sepals longer than the fruit. Seeds flattened, with hairy winged margins. HABITAT: fields, stony and sandy places; Spain to Palestine. April–July. Native in Jersey; a rare casual in Britain.

L. micrantha (Cav.) Spr. has the smallest flowers of all the Mediterranean toadflaxes, only 4–8 mm. in length including the spur; they are purple. Calyx with glandular hairs; seeds with a broad wing. HABITAT: cultivated ground, sands and rocks; circum-Medit. March–May.

SCROPHULARIA—Figwort

S. peregrina L. 178   NETTLE-LEAVED FIGWORT

An annual with paired, nettle-like leaves and dark brown-purple flowers in long-stemmed clusters in the axils of the leaves. Stem hollow, 30–60 cm. high, 4-angled, often reddish. Leaves light green, egg-shaped, pointed, heart-shaped or straight-edged at the base, strongly and unevenly toothed. Flowers in long-stalked clusters of 2–5; corolla 5–8 mm. long with a circular staminode (a small flap-like structure in the corolla in the place of the 5th stamen); sepals green without a membraneous border. HABITAT: cultivated ground, vineyards, bushy places, walls; European Medit. April–June.

S. lucida L.

A somewhat grey-green shining-leaved plant with deeply cut compound leaves and elongated leafless heads of small reddish brown flowers. A biennial with branched stem, 15–50 cm. high, and lower leaves simply lobed and upper leaves once or twice cut into oblong-toothed lobes. Flowers

169

all stalked, 4–9 mm. long, with the stamens more or less included in the corolla, staminode kidney-shaped. HABITAT: rocky and stony places, cliffs, walls; France to Palestine. March–June.

The similar *S. canina L.* is a perennial with smaller black-purple flowers with the margin of the upper petals white, and with projecting stamens. Flowers 3 mm., less commonly up to 6 mm. long; staminode acute lance-shaped, sometimes absent. Leaves very variable, once or twice cut. HABITAT: dry hills, rocks, screes; circum-Medit. March–July.

VERONICA—Speedwell
V. cymbalaria Bodard 403
Probably the most widespread annual speedwell of cultivated ground in the Mediterranean. The tiny white flowers are borne singly in the axils of the broad leaves. A creeping plant, 10–60 cm. long, with leaves all stalked, the upper alternate, blade broader than long with 5–11 rounded lobes, the terminal broadest. Flowers very small, scarcely longer than calyx, on long stalks longer than the leaves; calyx with 4 oval, blunt, hairy lobes narrowed at the base. Fruits hairy. HABITAT: cultivated places, walls, rocks and rough ground; circum-Medit. February–April.

BELLARDIA
B. trixago (L.) All. (Trixago apula Stev., Bartsia trixago L.) 175
A stiff erect glandular-hairy plant with short dense leafy spikes of attractive large whitish flowers flushed with pink or yellow. An annual, 10–80 cm. high, with oblong lance-shaped leaves with large widely spaced teeth. Flower spike dense with glandular bracts; corolla hairy, 2 cm. or more long, and 2–3 times the length of the calyx, with a broad lower lip and 2 bosses in the throat. HABITAT: sandy places, fields, especially where damp; circum-Medit. April–June.

PARENTUCELLIA
P. latifolia (L.) Car. (Bartsia latifolia S. & S.) 404
A low reddish, hairy, glandular annual with small opposite toothed leaves and small reddish-purple flowers which are 2-lipped, and with a whitish tube. Stem 5–20 cm. high, unbranched, or branched only from the base, with few stalkless oval leaves nearly as broad as long and deeply cut almost to the base. Flowers small, 1 cm. long, upper lip 3-lobed, lower 1-lobed, densely hairy outside, white within, tube long, white. Fruit hairless, or shortly hairy, equalling the calyx. HABITAT: dry hills, sandy places, pastures along the littoral; circum-Medit. March–June.

OROBANCHACEAE—Broomrape Family
A very distinctive family of plants without any green colour to the leaves or stems. They are parasites and their leaves are reduced to scales. Flowers

in axils of bracts in dense terminal spikes. Flowers symmetrical in one plane only, corolla 2-lipped, often with a curved tube; calyx tubular. Stamens 4, fused to corolla; ovary 1-celled; fruit a splitting capsule.

The Broomrapes are annual or perennial root-parasites with tuberous swellings attached to the roots of host plants, but it is often difficult to tell which plants are parasitized as the flowering spikes may arise some distance away. Broomrapes are parasites on many species of flowering plants particularly of the daisy, pea, carrot and mint families.

OROBANCHE

O. ramosa L. (*Phelypaea ramosa* (*L.*) *C.A. Mey.*) **180**

BRANCHED BROOMRAPE

Distinguished from many others by the flowering spike which is normally branched at the base into several lateral spikes; in poorly developed specimens the spike may be unbranched (as in the illustration). Flowers in a long loose spike; corolla small, 10–12 mm. long; commonly pale blue throughout but sometimes pale blue or whitish yellow tipped with blue-purple; upper lip of corolla with 2 rounded lobes, lower with 3 blunt lobes. Calyx with 4 pointed teeth shorter than calyx tube. Parasitic on potato, tomato, tobacco, hemp, and others. HABITAT: cultivated ground; circum-Medit. May–September. Introduced into Britain.

O. crenata Forskål **181**

A parasite on members of the pea family such as peas, beans, vetches, etc., forming a robust dense cylindrical spike of large flowers, usually whitish and streaked with blue-violet. Flowering stems 20–70 cm. high; corolla 2½–3 cm. long, more or less straight-tubed, lower lip with 3 very unequal lobes with the central lobe twice as broad as the laterals; stamens attached near the base of the corolla tube, filaments hairy at the base. Flowers scented of carnations and vary in colour from white to yellow with red, blue or purple veins. HABITAT: fields, uncultivated ground, gardens; circum-Medit. March–June.

The similar *O. gracilis Smith* has medium-sized flowers, 1½–2½ cm. long, with a regularly curved corolla tube which is yellowish red and blood red within, or flushed with violet; lower lip of corolla of 3 equal lobes. Stigma yellow, edged with purple. The whole plant is reddish, hairy and glandular, and it smells of clover. Parasitic on members of the pea family. HABITAT: fields, woods; European Medit., Turkey, N. Africa. April–July.

O. minor Sm.

LESSER BROOMRAPE

A very variable species with usually small yellowish flowers tinged and veined with purple, and stamens attached to the middle of the corolla tube, stigma usually violet-purple. Flowering stems 10–50 cm. Flowers not more than 1½ cm. long, with a regularly curved tube with the upper lip directed forwards as a continuation of the tube, not at an angle to the tube. Stalks of stamens almost hairless. A species with a number of distinct varieties living on different host plants. Parasitic on herbaceous members of the pea and daisy families. HABITAT: clover fields, meadows, at times in large numbers; circum-Medit. April–July. A British native.

GLOBULARIACEAE—Globularia Family

Shrubs or shrublets in our area, with alternate simple leaves without stipules. Flowers in rounded heads surrounded by an involucre of numerous bracts. Calyx 5-lobed; corolla fused, 1–2-lipped. Stamens 4, attached to the top of the corolla tube; anthers kidney-shaped; ovary 1-celled with 1 seed.

GLOBULARIA

G. alypum L. **179**

A low-growing branched under-shrub with brittle stems and leathery leaves, and globular heads of many tiny pure blue flowers. Up to 60 cm. high, with evergreen lance-shaped leaves with spiny tips and sometimes several spiny teeth. Flower heads sweet scented, 1½–2 cm. across, with bracts below heads brown-papery with hairy margins; calyx and silky hairs; corolla with a single 3-lobed lip. HABITAT: dry places and rocks usually in maquis; circum-Medit. Winter and spring.

The plant is a violent purgative.

ACANTHACEAE—Acanthus Family

Distinguished from the Scrophulariaceae by obscure botanical characters such as the hardening of the stalks of the ovules which often acts as a catapult mechanism throwing out the seeds. In addition, leaves always opposite, flower spike often with coloured bracts. Corolla usually 2-lipped.

ACANTHUS

A. mollis L. **397**   BEAR'S BREECH

A handsome and striking perennial with very large, deeply cut dark green leaves, and with a long terminal spike of very large white purple-veined, 3-lobed flowers, set among spiny bracts. Flowering spike up to 60 cm.; leaves stalked, soft and not spiny, deeply divided nearly to the midrib into large pointed-toothed lobes. Bracts subtending the flowers egg-shaped, with spiny teeth tinged with purple; calyx hairless; corolla of a single 3-lobed lip, 3–5 cm. long. HABITAT: hills, cool and rocky places; Spain to Greece, N. Africa. May–July.

The similar *A. spinosissimus Pers. (A. spinosus L.)* has stiff deeply dissected leaves with stiff sharp white spines. HABITAT: dry hills, derelict cultivation; Italy to Greece, Turkey. May–July.

The leaves of these and other Acanthus species were the motifs for the designs of the capitals of Corinthian columns of the Greeks and Romans. These plants are often cultivated.

172

PLANTAGINACEAE—Plantain Family

Annual or perennial herbs with usually all leaves basal and forming a rosette. Flowering stems usually leafless bearing close heads or spikes of small flowers with parts in 4's. Corolla fused, petals papery; calyx fused at the base, remaining around fruit; stamens 4 with long thin stalks and large anthers; ovary 1–4-celled. Fruit a capsule splitting transversely, or 1-seeded nutlet.

PLANTAGO—Plantain

P. indica L. (*P. psyllium* L.) 182

A plantain with an erect branched stem which bears leaves along its length and a group of rounded terminal long-stalked flower heads. An annual with hairy and sticky stems which are sparsely branched, 10–35 cm. high. Leaves opposite or in groups, stalkless, strap-shaped and flat, 3–6 cm. long. Bracts subtending flower heads narrow and not as long as flower-stalks. Flower heads many, ovoid, 6–13 mm. long, glandular-hairy. HABITAT: grassy places, fields, sandy places, track sides; circum-Medit. April–June. A British casual.

P. lagopus L. 405

A plantain with very silky ovoid or hemispherical heads of flowers borne on slender stems, and narrow leaves much shorter than the flower stems. A rosette annual with many erect green and sparsely hairy lance-shaped leaves, each with 3–5 veins, and long stalks. Flower stalks several, 10–40 cm. long, with shallow furrows, each bearing a single compact head, 2 cm. by 1 cm. Corolla, sepals and bracts all silvery-haired. HABITAT: dry sandy places, fields, track sides and uncultivated places; circum-Medit. March–June.

RUBIACEAE—Bedstraw Family

Woody plants or herbs with whorled or opposite leaves and with stipules. Flowers small, in clusters, with symmetrical corollas and parts in 4 or 5's. Corolla funnel-shaped with spreading petals; stamens attached to the corolla tube. Ovary inferior, 2-celled; fruit a capsule, berry or splitting into 2 one-seeded units.

PUTORIA: leaves opposite, stipules very small and differing from the leaves; corolla with long tube.

RUBIA: leaves in whorls and stipules leaf-like; corolla with very short tube and spreading petals.

PUTORIA

P. calabrica (*L. fil.*) *DC.* 183

A fetid plant with small narrow leaves with inrolled margins and rosy-red terminal clusters of flowers with long tubes and reflexed petals. Stems woody at the base and tufted, 10–30 cm. high, with opposite narrow-lance-shaped or linear leaves, about 1 cm. long. Flowers with calyx 2–3 mm.

long; corolla tube about 1½ cm. long; stamens 4 with red anthers. Fruit fleshy of 2 units. HABITAT: rocky and gravelly places on limestone; circum-Medit. (not France). April–September.

RUBIA
R. peregrina L.   WILD MADDER
An evergreen perennial with trailing, scrambling, quadrangular stems, whorls of 4–6 dark green prickly leaves and clusters of small greenish-yellow flowers. Stem woody at the base, 30–120 cm. long, sharply 4-angled and with downward pointing spines; leaves narrow-oval, 1½–6 cm. long, rigid, leathery, shining above, with horny toothed margins. Flowers many in a terminal or axillary cluster; corolla 5 mm. across, with pointed lobes; fruit black, 4–6 mm. HABITAT: woods, bushy places, hedges and rocky places, largely on limestone; circum-Medit. May–July. A British native.

CAPRIFOLIACEAE—Honeysuckle Family

Mostly shrubs with soft wood and wide pith, with opposite leaves, and stipules mostly absent. Flowers usually in flat-topped clusters with parts of flower in 5's, except the ovary. Corolla fused into a tube with the stamens attached to it. Ovary inferior, carpels 2–5; fruit a drupe, berry or 1-seeded unit. A family of many ornamental species.

VIBURNUM: flowers symmetrical in compound, umbel-like heads; ovary 1-celled with 1 ovule; fruit with a hard stone.

LONICERA: flowers strongly 2-lipped with a 4-lobed upper lip and a single lower lobe, or flowers regularly lobed; ovary 2-celled with numerous ovules; fruit a few-seeded fleshy berry.

VIBURNUM
V. tinus L. **186**   LAURESTINUS
A large shrub or small tree with dark green smooth laurel-like leaves and flat heads of small white or pink flowers. Trunks with grey bark, up to 3 m. high, young branches hairy, often reddish. Leaves leathery, evergreen, shining above and hairy-glandular on the underside, oval with untoothed margins. Flowers in a close flat cluster, faintly scented; corolla tubular with 5 rounded petals and projecting stamens. Fruits globular, turning metallic bluish-black at maturity. HABITAT: woods, thickets, garigue, and sheltered valleys near the sea; European Medit., Lebanon, N. Africa. February–June.
Commonly cultivated for its winter flowers in Britain. The fruits are a rather drastic remedy for dropsy.

LONICERA—Honeysuckle
L. etrusca Santi
A climbing honeysuckle with the upper pairs of leaves fused together,

174

and with flowers in long-stalked, many-flowered clusters. Climbing plant up to 4 m. high, with somewhat leathery deciduous leaves, 3–8 cm. by 2–6 cm., oval and blunt, the lower more or less stalkless, the upper encircling the stem. Flower clusters in groups of 3 with additional lower clusters; stalks of flower clusters 3–4 cm. long; clusters often of 12 flowers or more. Corolla 3–4 cm. long, creamy coloured and flushed with red outside, slightly scented. Fruit ovoid, red. HABITAT: woods, hedges, thickets, always on limestone; circum-Medit. May–July.

L. implexa Ait. **406**

A climbing honeysuckle with evergreen leathery leaves and with stalkless terminal flower clusters surrounded by cup-shaped encircling bracts. A woody climber up to 2 m., with evergreen more or less oval hairy leaves with narrow transparent margins, the upper ones mostly fused together and encircling the stem. Flower clusters about 6-flowered and corolla cream coloured, reddish on the outside, 3–4½ cm. long, the interior of corolla tube and style often hairy. HABITAT: woods, hedges, rocks; circum-Medit. April–June.

VALERIANACEAE—Valerian Family

Herbaceous plants often with strongly smelling underground parts. Flowers often in a head of many small flowers; corolla with a thin spur or pouch-like sac at the base. Calyx variously toothed, often inrolled in flower and forming a feathery pappus in fruit. Stamens 1–4, attached to the base of the corolla tube. Ovary inferior, 3-celled, but in fruit only 1 cell develops a seed.

VALERIANELLA: annuals with tiny flowers in terminal heads; corolla funnel-shaped; calyx forming a tooth or funnel-shaped rim; stamens 3.

CENTRANTHUS (or KENTRANTHUS): perennials with calyx inrolled in flower and enlarging to form a pappus in fruit; corolla tube with spur or sac at base; stamen 1.

FEDIA: an annual with long-tubed flowers and 2-lipped petals; stamens 2 or 3; fruit of 2 kinds.

VALERIANELLA—Corn-Salad

V. echinata (L.) DC.

A small annual corn-salad with few branches, light green leaves and with small rounded terminal heads of tiny pink or lilac flowers. Stems thick, particularly under the fruits, 5–20 cm. high, with oblong-obovate, toothed leaves. Fruiting heads generally in pairs, very distinctive and recalling a miniature knobkerrie with its cluster of curved spines on a thickened stalk. Each fruit has 3 spines with one much stouter and longer than the others. HABITAT: fields; circum-Medit., not Morocco. April–May.

V. vesicaria (L.) Moench **(407)** is similar but the fruits are unlike those of any other species in being swollen and bladder-like. A sparsely branched, slightly hairy annual with lance-shaped leaves. Flowers in globular paired stalked heads; fruits about 5–10 on each head. Individual fruits globular, hairy, about 5 mm. in diameter, formed by the inflated calyx and with a circular apical cavity furnished with 6 small triangular incurved teeth with their points meeting in the centre. HABITAT: fields and derelict cultivation; France to Palestine. March–June.

CENTRANTHUS (Kentranthus)

C. ruber (L.) DC. **184**   RED VALERIAN

A hairless bluish-grey perennial with tufted growth and many erect stems bearing close terminal clusters of small, usually deep rose coloured flowers. It has stout herbaceous stems up to 80 cm. high growing from a thick sweet-scented rootstock. Leaves about 10 cm. long, egg-shaped to lance-shaped with entire margins, the upper toothed and stalkless. Flowers in rounded flat-topped sweet-scented heads, usually rose but occasionally white; corolla about 5 mm. across with a slender tube, 8–10 mm. long, and a narrow spur twice as long as the ovary; the single stamen projecting beyond the corolla. HABITAT: rocks, cliffs and old walls; circum-Medit. March–September. Naturalized in Britain.

The seeds were used in ancient embalming.

C. calcitrapa DC.

An annual, often reddish in colour, with tiny pink flowers ranged in 2 ranks on one side of the diverging stems and forming a spreading terminal head. Plant 10–40 cm. high with simple erect stem which is hollow and finely ridged. Lower leaves uncut, upper deeply cut into narrow segments which are hairless. Flowers very small, 2–5 mm. long, with the spur reduced to a boss, usually pink but sometimes white. HABITAT: dry hills, rocks; European Medit., Cyprus, Lebanon, N. Africa. April–July.

FEDIA

F. cornucopiae (L.) Gaertner **185**

A valerian-like plant with thick hollow stems branched dichotomously and terminating in paired clusters of rose coloured flowers on thick stalks. A hairless annual, 10–30 cm. high, with oval-elliptic leaves, sparsely toothed at the base. Flowers in stalkless clusters; calyx minute; corolla with a long narrow tube, pouched at the base, 2-lipped and with 5 unequal lobed petals. Stamens 2 or 3; fruit of 2 kinds, the central fertile, the outer sterile. HABITAT: cultivated places, pastures, rocks and sands; European Medit. (not France), Cyprus, N. Africa. March–June.

DIPSACACEAE—Scabious Family

Usually herbaceous plants with opposite leaves. Flower heads dense, of many florets surrounded by a calyx-like ring of bracts (involucre). Florets surrounded by an extra calyx; calyx cup-shaped or deeply cut into 4–5 segments, or into numerous teeth or hairs. Corolla fused, often forming a curved tube, 2-lipped with 4–5 similar lobes; stamens usually 4. Ovary inferior, 1-celled.

KNAUTIA: receptacle to which flowers are attached hemispherical, hairy, without bracts; calyx of 8 teeth or bristles.

SCABIOSA: receptacle elongated, not hairy but with lance-shaped bracts; calyx with bristles or teeth.

KNAUTIA
K. integrifolia (L.) Bert.
An annual scabious with a basal rosette of deeply cut leaves and narrow stem leaves, and with medium-sized pale pink or lilac flower heads. A hairy plant, 30–60 cm. high, the basal leaves with egg-shaped segments, the upper stem leaves cut into narrow linear segments, or uncut and narrow lance-shaped. Flower heads surrounded by 8–12 hairy, lance-shaped and long-pointed bracts, shorter than the flowers; flower heads almost flat with outer florets spreading. Fruit surmounted by two sets of teeth and with numerous white hairs. A very variable species. HABITAT: fields, hills and rocky places; European and Asiatic Medit. April–June.

SCABIOSA—Scabious
S. atropurpurea L. var. maritima (Torn.) Beg.
A pink or lilac flowered scabious with distinctive ovoid or cylindrical fruiting heads with many long russet-coloured spines formed from the calyx. A branched annual or perennial, up to 1 m. high, with leaves toothed or cut into narrow linear segments. Flowering head flat, with outer florets spreading with 5 unequal lobes. Fruiting head greatly elongating and becoming cylindrical; fruit hairy with 8 ribs, and with a very narrow membraneous infolded crown round the top; calyx long-stalked, with 5 long spines. HABITAT: sands and uncultivated ground, commonly by the sea; circum-Medit. June–October.

S. stellata L. 408
A blue flowered scabious which is very distinctive in fruit, for the individual fruits have broad yellowish papery frills, and together they form a globular fruiting head 2–3 cm. across. A rough hairy annual, 10–40 cm. high, with basal leaves toothed or deeply cut, and stem leaves divided into long narrow pointed segments. Flowering head blue, 2–3½ cm. across, which after flowering becomes spherical and entirely covered with large membraneous yellow crowns 1 cm. or more across. Calyx with 5 long bristles as long or longer than the crowns. A very variable plant. HABITAT: hills and uncultivated ground; Spain, France, Sardinia, Dalmatia, Turkey, N. Africa. April–July.

S. prolifera L. 187   CARMEL DAISY
A striking scabious with large cream-coloured flower heads and globular

fruiting heads covered in papery cup-shaped frills. A stout annual, 30–60 cm. high, with forked stems; leaves large, oblong lance-shaped, very hairy, with untoothed margins. Flower heads large, 3–4 cm. across, often stalkless, in the angles of the branches; outer florets with long spreading cream-coloured petals. Fruiting head of many rust-coloured papery ribbed crowns each up to 1½ cm. across, with 30–34 nerves; calyx with 5 long bristles. HABITAT: fallow fields on clay and limestone; Turkey to Palestine, Cyprus, Libya. March–May.

CUCURBITACEAE—Cucumber Family

A family of rough-hairy herbaceous plants which usually climb or scramble by means of tendrils which are spirally coiled. Flowers 1-sexed, usually with sepals, petals and stamens in 5's, and an inferior ovary which develops into a fleshy, in our species elastically swollen or gourd-like fruit. Largely a tropical or sub-tropical family which produces a number of useful plants such as melons, cucumbers, gourds and calabashes.

ECBALLIUM

E. elaterium (L.) Rich. **188**   SQUIRTING CUCUMBER
A poisonous rough shaggy cucumber-like plant with small yellow flowers and green swollen hairy fruits which are easily detached from their stalks and explode their seeds outwards to some distance when disturbed. A spreading perennial with thick fleshy stems, 20–60 cm. long, without tendrils. Leaves thick, hairy, triangular with heart-shaped bases, irregularly lobed and toothed, whitish underneath. Flowers in axillary clusters, either male or female, the latter usually solitary. Fruit bent downwards when ripe, 4–5 cm. long, oblong-cylindrical, rough and green. HABITAT: track sides, rubbish dumps and waste land; circum-Medit. March–September.

The root contains a violent purgative and is used for rheumatism, paralysis, dropsy and shingles.

CAMPANULACEAE—Bellflower Family

Mostly herbaceous plants, very commonly with milky juice. Leaves usually alternate, simple, without stipules. Flowers usually large and bright; calyx tube fused to ovary; corolla bell-shaped or tubular with 5 equal lobes; stamens 5 attached to base of the corolla. Ovary with 3–5 styles and 2–10 cells; fruit fleshy or a splitting capsule.

CAMPANULA: fruit ovoid or top-shaped with a persistent calyx and splitting by pores or valves; corolla bell- or funnel-shaped.

LEGOUSIA: (sometimes spelt *Legouzia*): similar to *Campanula* but fruit much elongated; tube of corolla not bell-shaped, petals spreading.

178

CAMPANULA—Bellflower

C. rupestris S. & S. (including **C. celsii** DC., **C. tomentosa** Vent. and **C. andrewsii** DC.) **189, 190**

A very handsome campanula of cliffs and walls with spreading prostrate stems and silky grey leaves, bearing beautiful blue bells ranged individually but freely along the stems. A short-lived perennial forming a basal rosette of grey, hairy, egg-shaped and deeply lobed, toothed leaves, with flowering stems spread outwards from the old rosette, up to 30 cm. long. Flowers borne singly on short stalks in the axils of small egg-shaped leaves; corolla bell-shaped, about 2 cm. long, lavender-blue to intense blue; calyx shorter than corolla with triangular lance-shaped teeth.

A very variable species with many varieties and forms, of which *ssp. anchusaeflora* S. & S. (189) is remarkable for its long tubular deep violet-blue flowers. HABITAT: rocks and cliffs; Greece and the Islands. April–July.

C. drabifolia S. & S. **193**

A small weak hairy annual with rough oval leaves and small blue flowers. Stems dichotomously branched with spreading hairs and hairy leaves which are more or less toothed. Flowers shortly stalked in a loose spreading cluster, blue-violet in colour with a white throat; corolla bell-shaped, 10 mm. long, divided to ⅓ of its length into 5 teeth. Calyx ½ the length of the corolla but enlarging and star-shaped in fruit. HABITAT: rocks and dry places; Greece and the Islands. April–June.

The similar **C. erinus** L. is also a weak spreading annual but it has much smaller flowers which are as long as or a little longer than the calyx. Flowers 3–5 mm. long, white or pale reddish-lilac, and like the previous species the calyx enlarges and becomes star-shaped in fruit. HABITAT: walls, grassy and stony places, largely on limestone; circum-Medit. April–June.

C. spathulata S. & S. *ssp.* **spruneriana** (*Hampe*) *Hay.*

Somewhat like our harebell but with larger more funnel-shaped flowers, and calyx with long narrow teeth. A slender perennial with tall widely spreading branches with few leaves, up to ½ m. high. Leaves hairless or sparsely hairy, the lower oblong-lance-shaped with shallow rounded teeth and narrowed into a leaf stalk, the upper narrow lance-shaped, stalkless. Flowers solitary on very long stalks, corolla blue, funnel-shaped, 2½–3 cm. long; calyx teeth very narrow lance-shaped and ending in a long point nearly as long as the corolla. A very variable plant. HABITAT: in woods and bushy places; Greece and the Islands. April–June.

C. rapunculus L. RAMPION

A narrow-leaved campanula with a narrow elongated spike of many short-stalked flowers. A slender-stemmed, sparsely branched, hairy, somewhat rough biennial, up to 80 cm. high; leaves few, lower oblong and bluntly lobed, broader towards the apex and narrowed into a leaf-stalk; upper lance-shaped, stalkless. Flowers erect, 2–2½ cm. long, broadly

179

funnel-shaped and narrowed at the base, blue-lilac with triangular lobes ⅓ the length of the tube; calyx teeth narrow lance-shaped, much shorter than the corolla. HABITAT: fields and bushy places, vineyards and track sides; circum-Medit. May–July. Introduced into Britain.

C. ramosissima S. & S.   191

An annual bellflower with large open violet flowers on slender stems. 30–50 cm. tall, rarely branched at the base, with hairy, toothed, oval-lanceolate leaves becoming narrower upwards. Flowers 3 cm. across, with prominent style, and narrow sepals visible between deeply cut, broad petal lobes. HABITAT: grassy and stony places, olive groves; Italy to Greece, Palestine. April–May.

LEGOUSIA

L. speculum-veneris (L.) Fisch. (Specularia speculum (L.) DC.)   192

VENUS'S LOOKING-GLASS

An erect, slender, sparsely branched annual with widely opening bright violet-blue flowers. Up to 40 cm. high with stalkless oblong leaves with feebly waved margins, hairless or sparsely hairy. Flowers stalkless, some solitary, others in clusters of 3–5 forming a lax spike; corolla 10–12 mm. long, with petals as long as the calyx teeth which are linear, long-pointed and spreading. Fruit 10–15 mm. long, contracted at the apex. HABITAT: cultivated ground, fields and waste places; European Medit. to Palestine, Egypt. April–July.

The similar L. pentagonia (L.) Thell. is a rough-hairy plant with larger flowers, 15–18 mm. long and 2–3 cm. across, spreading into a pentagonal star. Calyx covered with long white hairs; teeth short, 2–3 times shorter than tube, spreading in fruit. Fruit not narrowed at apex. HABITAT: corn-fields and waste places; European Medit. to Palestine. April–July.

COMPOSITAE—Daisy Family

The largest and most cosmopolitan family of the flowering plants and probably the most ubiquitous in the Mediterranean. Mostly herbaceous plants with milky juice or oil ducts, and leaves without stipules. Flowers arranged in a characteristic head (*capitulum*) composed of many small complete flowers (*florets*), and surrounded closely by bracts (*involucre*). The florets and involucre are attached to the swollen end of the stalk, the *receptacle*. The florets may all be similar, or of 2 kinds, the inner *disk-florets* with short tubular corollas and the outer *ray-florets* with long strap-shaped corollas.

The family is divided into 2 main sub-sections: *Tubuliflorae* with some or all florets disk-florets, and the *Liguliflorae* with all the florets ray-florets, as in the dandelion. Calyx either forming a *pappus* of numerous simple or feathery hairs, or a row of membraneous scales or teeth. Stamens 5 fused by their anthers; ovary inferior 1-seeded: fruit with or without a pappus.

There are many edible members of the family, e.g. lettuce, endive, salsify, Jerusalem and globe artichokes and chicory; some are medicinal, many are ornamental.

BELLIS: leaves usually basal, heads solitary with a single row of white or pink ray-florets; pappus absent.

EVAX: flower heads practically stalkless and surrounded by many involucral bracts, the whole forming a rosette and appearing as a single head.

HELICHRYSUM (*ELICHRYSUM*): leaves white-cottony with borders inrolled; involucre often highly coloured and papery; upper bracts blunt-tipped.

PHAGNALON: leaves white-cottony with borders inrolled; heads solitary; involucre papery or leathery, bracts acute.

PALLENIS: heads yellow with both disk- and ray-florets; involucre of 2–3 ranks with the outer bracts flattened and ending in long sharp spines.

ODONTOSPERMUM: heads yellow with disk-florets and 1–2 ranks of ray-florets; involucre of several ranks, bracts all or most leafy.

XANTHIUM: flowers small and inconspicuous, the male only in heads, the female below, 1–2 flowered; fruit hard, spiny.

ANTHEMIS: flower heads with disk-florets and generally white ray-florets; receptacle with scales; leaves 1–3 times cut into narrow segments; fruit without pappus.

CLADANTHUS: a genus of one species only, similar to *Anthemis* but typically producing short branches from immediately below the solitary yellow-rayed flower head. Leaves feathery.

OTANTHUS: involucre of numerous overlapping woolly bracts; florets all disk-florets and corolla-tube prolonged downwards into 2 ear-like spurs almost enclosing the ovary.

CHRYSANTHEMUM: receptacle flat without scales; ray-florets white or yellow; fruit ribbed, without hairy pappus.

SENECIO: involucre in 1 rank or with a few shorter bracts; receptacle without scales; ray-florets usually present; fruit with a pappus of simple hairs.

CALENDULA: involucre in 1 rank; ray-florets in 2–3 ranks, orange or yellow; fruits large, curved or ringed, smooth or prickly on the back.

ECHINOPS: leaves and stem spiny, thistle-like; flower head spherical, of many spreading 1-flowered capitula or 'headlets' with spiny bracts.

CARLINA: leaves and stem spiny; outer involucral bracts spiny, leaf-like, inner bracts papery coloured and spreading when dry.

ATRACTYLIS: involucre broadly bell-shaped or cylindrical of several ranks, the outer leaf-like and deeply cut into spiny teeth, the inner with papery tips and not spreading.

181

JURINEA: involucre with spiny tips, inner bracts longer; pappus feathery, the inner longer; florets all disk-florets, purple.

CARDUUS: thistles with pink or purple flowers with involucral bracts, usually spiny; pappus rough, unbranched.

NOTOBASIS: involucre spiny-tipped; heads large, purple, with outer florets sterile; pappus of inner florets feathery, pappus of outer florets with unbranched hairs.

SILYBUM: flowers large, solitary, purple; involucral bracts fringed with teeth and contracted into a leafy tip with a long spiny point.

GALACTITES: flowers with larger sterile outer florets; anthers with a curved appendage and fused stalks; involucre spiny; pappus feathery.

CRUPINA: heads narrow, cylindrical, long-stalked; involucre not spiny; pappus of 2 rows, outer of several rows of rough rigid spines, inner of 5–10 broad scales.

LEUZEA: heads large, solitary, reminiscent of a pine-cone; involucral bracts enlarged above into a rounded papery blade.

CENTAUREA: involucral bracts with papery, often fringed margins or with branched spines; outer florets spreading, sterile; pappus absent or of unbranched hairs.

CARTHAMUS: outer involucral bracts very similar to the spiny leaves, inner rounded, blunt, with spiny tip and papery, cut margins.

SCOLYMUS: involucre in several ranks, outer very like the spiny leaves; all florets strap-shaped ray-florets.

CICHORIUM: flowers blue, involucre of 2 ranks; all florets strap-shaped ray-florets.

TOLPIS: involucre of 1–2 ranks, very narrow and spreading outwards; all florets strap-shaped ray-florets; flowers yellow.

RHAGADIOLUS: involucre of 2 ranks, the inner of 5–8 cylindrical scales wholly enclosing the fruits, hardening and spreading outwards into a star in fruit; all florets strap-shaped ray-florets.

HEDYPNOIS: involucral bracts hardening and enveloping the ripe fruits; all florets strap-shaped ray-florets.

UROSPERMUM: involucre bell-shaped, bracts 8–10 in 1 rank, united below; all florets strap-shaped ray-florets.

TRAGOPOGON: leaves garlic-like, linear lance-shaped; involucral bracts united below and reflexed at the tip at maturity; fruit with a long beak surmounted by a spreading feathery pappus; all florets strap-shaped ray-florets.

ANDRYALA: annuals with whitish or rusty woolly hairs; involucre 1-ranked with few additional bracts; fruit tapering at base, 10-ribbed; pappus hairy, stiff, brown; all florets strap-shaped ray-florets.

182

LAUNAEA: spiny-stemmed and with interwoven branches; fruits of 2 forms, outer cylindrical angled, velvety or transversely wrinkled, inner white smooth; pappus soft, smooth; all florets strap-shaped ray-florets.

CREPIS: fruits cylindrical, tapering or long-beaked; pappus of soft white unbranched hairs; involucre 1-ranked; all florets strap-shaped ray-florets.

BELLIS

B. silvestris Cyr.

A much more robust plant than our common daisy and distinguished from it by the leaves which are gradually narrowed into a broad-winged leaf stalk, and by the 2–4 veins of the blade. A perennial with leafless flowering stems, 15–30 cm. high, and leaves all basal, the stalk longer than the oblong-oboval blade, with almost velvety grey hairs. Flower heads generally 3–4 cm. across, outer florets ray-florets and tinged with purple. HABITAT: fields, thickets and screes; circum-Medit. October–March.

B. perennis L., the COMMON DAISY, is very similar but its leaves have a single vein and the blade is suddenly contracted into the leaf stalk; flower heads 1–3 cm. across. HABITAT: fields, open woods, track sides; European Medit. October–May. A British native.

B. annua L. is an annual with a branched flowering stalk, usually bearing several leaves. Flowers small, 1–1½ cm. across. HABITAT: saline ground, damp and grassy places; circum-Medit. March–June.

EVAX

E. pygmaea (L.) Pers. (Filago pygmaea L.) **409**

A tiny white-woolly annual pressed against the ground with leaves forming a rosette encircling stalkless yellow flower heads. An annual 1–4 cm. high, with blunt oblong-obovate stalkless leaves which are densely covered in white woolly hairs and are closely bunched and overlapping. Flower heads very compact in the centre of the rosette, with long pointed yellow bracts. Fruits covered with rough swellings. HABITAT: sands, grass, garigue near the sea; circum-Medit. April–May.

HELICHRYSUM—Everlasting

H. siculum (Spreng.) Boiss.

A low growing spreading white-leaved plant with many erect stems bearing narrow, woolly leaves, and globular heads of shining bright yellow 'everlasting' flowers. A perennial with a woody base with stems up to 30 cm. high. Leaves linear, about 3 cm. long, inrolled at the edges, greyish above with cobweb-like hairs, and white-woolly below. Flower clusters somewhat spreading to 4 cm. across, borne on sparsely leaved stems, with the individual heads 4–6 mm. long, and with lemon-yellow, glossy, thin, egg-shaped overlapping bracts. Outermost bracts 2–3 times shorter than the inner which are narrower and hairy on the back. HABITAT: stony, rocky places; Greece to Palestine. April–August.

The very similar *H. stoechas* (*L.*) *DC.* (**195**) is at once distinguished by its aromatic curry-like smell when crushed. Flower clusters densely globular, rarely spreading. Heads smaller, external florets female only. HABITAT: stony, rocky, sandy places; Spain to Italy, Morocco, Algeria. April–June.

H. sanguineum (*L.*) *Kostel*   194

A striking 'everlasting' with rounded heads surrounded by deep crimson bracts, which grows in the Eastern Asiatic Mediterranean region. A perennial with an erect stem branching only in the upper part; lower leaves 5–10 cm. long, oblong blunt, upper leaves shorter, narrow lance-shaped, with blades clasping and running some way down the stem; stem and leaves thickly covered with white woolly hairs. Flower clusters forming dense rounded heads, about 1½ cm. across, with the individual heads small, 3–4 mm. long. Involucral bracts rounded, blunt, thin and deep crimson, contrasting with the white woolly hairs of the flower stalks. HABITAT: hills and shrubby places; Turkey to Palestine. April–May.

PHAGNALON

P. rupestre (*L.*) *DC.*   196

A small undershrub with intensely white-woolly stems and many narrow wavy-edged leaves, and ultimate branches leafless and bearing single shining brownish-yellow flower heads. A compact rather bushy plant up to 30 cm. high, with narrow-lance-shaped pointed leaves with widely spaced teeth, green above, white below; upper leaves with inrolled margins. Flower heads ovoid, about 1 cm. long, with many closely pressed shining, smooth, stiff, dry bracts; outer bracts egg-shaped, blunt, inner broad-oblong, blunt. HABITAT: rocks, stony places, gravel; circum-Medit. April–June.

The *ssp. graecum Boiss. & Heldr.* has the involucre with linear, acute inner bracts, and the outer ones are triangular lance-shaped, narrowed to a slightly blunted point. It is common in Greece.

PALLENIS

P. spinosa (*L.*) *Cass.* (*Asteriscus spinosus Schultz-Bip.*)   202

A yellow-flowered composite with formidable-looking green bracts tipped with sharp spines spreading from underneath the flower head. A hairy annual or biennial, ½–1 m. high, branched from the base. Lower leaves oblong, broader towards tip and narrowed and elongated into a leaf stalk, upper leaves lance-shaped, stalkless. Flower heads about 2 cm. across; outer bracts rigid-spined, and spreading into a star to twice the width of the head; inner bracts egg-shaped, tough, papery and hairy. HABITAT: rocky places, olive orchards, fields and waysides; circum-Medit. April–June.

184

ODONTOSPERMUM

O. maritimum Schultz-Bip. (*Asteriscus maritimus* (*L.*) *Less.*) **197**

A soft shrublet with many stems forming mats with many hairy spatula-shaped leaves and large deep yellow marigold-like flowers surrounded by leafy bracts. Stem hairy, leaves stalked and obovate. Flower heads terminal, up to 4 cm. across, with many disk- and about 30 ray-florets, with narrow strap-shaped limbs, 1½ cm. long, and finely toothed at the end. HABITAT: rocky places by the sea and dry places inland; European Medit., N. Africa. March–June.

XANTHIUM

X. spinosum L. **410**   SPINY CLOTBUR

An extremely spiny South American plant with small heads of green flowers and spiny burred fruits. A hairy annual, 30–60 cm. high, branched from the base and bearing long yellow needle-like 3-branched spines, 1–3 cm. long, at the base of the leaves. Leaves soft, white-hairy underneath, lance-shaped, undivided or with 3–5 sharp lobes. Flowers in small rounded heads in the axils of the leaves; fruit green, hairy, oblong-cylindrical, 10–15 mm. long, bristling with narrow hooked yellow spines. HABITAT: waste places, track sides, banks and hedges; now circum-Medit. July–September. A British casual.

Not the 'thorns' of the Bible as suggested by some, in view of the plant's South American origin. First introduced to Portugal and now spread throughout the Mediterranean.

ANTHEMIS

A. tinctoria L.   YELLOW CHAMOMILE

Distinguished from other chamomiles by the yellow ray-florets (white in other species). An erect branched perennial plant, 30–60 cm. high, with dark green foliage. Leaves compound, divided into narrow lance-shaped segments which have saw-toothed margins. Flower heads large, 2½–4 cm. across, golden yellow, with the outer ray-florets rather short and narrow, 6–10 mm. by 2 mm. HABITAT: sunny slopes, rocks, railway tracks, walls, largely on limestone; European and Asian Medit. May–August. A casual in Britain.

The flowers yield a yellow dye.

A. chia L. **203**

One of the commonest small chamomiles which occur in great abundance in the spring in the Eastern Mediterranean. A low erect or spreading annual, up to 30 cm. high, with leaves egg-shaped in outline but twice cut into narrow pointed spreading segments which are further cut. Flower heads 3 cm. across, with white ray-florets over 1 cm. long and longer than the width of the disk; involucral bracts hairless with a rusty membraneous margin. Fruits cylindrical and ribbed, the outer ones somewhat curved, with a translucent projection as long as the fruit, the inner with a short projection or crown. HABITAT: fallow fields, dry, stony hillsides; Dalmatia to Palestine, Egypt. January–May.

A. tomentosa (L.) Urv.
A white-woolly, usually prostrate, spreading chamomile, with flower heads 2–3 cm. across, with the central orange disk broader than the white surrounding ray-florets. Leaves and stem covered with long woolly hairs, leaves twice cut with the segments rather blunter or more egg-shaped than in the preceding species. Bracts of flower heads woolly-haired. Fruits blunt, squarish in section with an obliquely cut-off apex, or with a short projection. HABITAT: sandy places by the sea and exposed hillsides; Greece and the Islands, Turkey. April–June.

OTANTHUS

Otanthus maritimus (L.) Hoffm. & Link (Diotis maritima (L.) Sm.,
   D. candidissima Desf.) 411   COTTON WEED
A silvery white spreading plant with numerous leaves and clusters of yellow flower heads covered in a felt of thick hairs. A stout-stemmed perennial, 10–40 cm. long, with many closely bunched spreading oblong leaves. Flower heads grouped at the ends of the branches in short, flattened heads; individual heads globose, 8–10 mm. across, of small yellow disk-florets (ray-florets absent) surrounded by an involucre of oval white woolly bracts. HABITAT: sands by the sea; circum-Medit. May–August. A very rare British native.

CLADANTHUS

C. arabicus (L.) Cass. 201
An annual reminiscent of a french marigold with a terminal head of large bright yellow flowers and fine much-divided leaves. Stem unbranched, about 15 cm. high, with leaves 1–2 times cut into thread-like segments about 1 cm. long. Flower heads about 3 cm. across, often with broad notched ray-florets and a small disk, with radiating branches arising directly below the heads and curving upwards; bracts papery with woolly hairs. HABITAT: cultivated places, dry pastures; S. Spain, N. Africa. March–June.

CHRYSANTHEMUM
C. segetum L. 198   CORN MARIGOLD
A smooth rather waxy-looking annual with lance-shaped bluish leaves and bright orange-yellow daisy-like flowers, 2½–4 cm. across. A sparsely branched plant, up to 50 cm. high, with lower leaves deeply cut and narrowed to a stalk, and the upper sparsely toothed and half encircling the stem. Stalks of flower heads thickened above; involucral bracts broadly egg-shaped with broad pale brown membraneous margins. Outer ray-florets orange-yellow, disk-florets the same colour. HABITAT: fields and

cultivated ground, largely on siliceous soils; circum-Medit. April–August. Probably introduced into Britain, but often a troublesome cornfield weed.

C. coronarium L. 199, 200 CROWN DAISY

A handsome showy plant with large golden-yellow flower heads and finely divided feathery leaves. A robust hairless annual growing up to 80 cm. high in years of good rainfall, with erect, branched, leafy stems. All leaves twice cut into sharply toothed lance-shaped lobes, the upper stalkless and half clasping the stem. Flower heads, up to 6 cm. across, borne on swollen stalks, with yellow disk-florets and similar coloured ray-florets with obovate strap-shaped limbs or sometimes with a paler central zone. *Var. discolor Urv.* (200) has bicoloured ray-florets. HABITAT: fallow fields and fields under crops, uncultivated ground; circum-Medit. March–June. Often cultivated in gardens. In the spring, in the Eastern Mediterranean, this plant grows in such abundance that fallow fields and waste ground are sometimes a sheet of brilliant yellow.

SENECIO

S. cineraria DC. (Cineraria maritima L.)

A striking shrubby plant with white-felted branches and segmented white-felted leaves, and branched flattish heads of yellow flowers with white woolly bracts. Up to 60 cm. high with leaves deeply cut to the midrib with segments further cut into oblong rather blunt lobes. Individual flower heads 8–12 mm. across, with 10–12 bright yellow ray-florets, and with very woolly oblong lance-shaped bracts; fruits hairless. HABITAT: sands and rocks by the sea; Spain to Dalmatia, Rhodes, Morocco to Algeria. May–August. Introduced into Britain. Often cultivated as an ornamental plant.

S. vernalis W. & K.

An annual groundsel with relatively large conspicuous yellow flowers. Stems usually branched from the base, 20–30 cm. high, and with deeply toothed and cut oblong leaves. Leaves and stems covered with sparse white cobweb hairs. Flower heads few, in spreading clusters, each head 2–3 cm. across with yellow spreading ray-florets nearly 1 cm. long. HABITAT: broken ground, fields, dry places by the sea, often very common. France to Palestine, N. Africa. A native of the Eastern Mediterranean but advancing westwards. November–May.

The very similar *S. coronopifolius Desf.* is distinguished by the stems and leaves being practically hairless, and by smaller flower heads, 12–15 mm. across. The leaves are all more deeply cut into long narrow segments. HABITAT: between fields, walls; Greece and the Islands; N. Africa. March–May.

CALENDULA

C. arvensis L. 204 FIELD MARIGOLD

The small-flowered orange marigold is a common plant of the Mediterranean, and its strange-looking sickle-shaped fruits set with stiff bristles

187

are unmistakable. It is an erect or spreading annual, up to 30 cm. high, with hairy branches. Leaves oblong lance-shaped, sparingly toothed or without teeth, stalkless and half embracing the stem. Flower heads solitary, 1–2 cm. across, and borne at the ends of the stems with the ray-florets twice as long as the involucral bracts. Fruits of 3 types: the outer sickle-shaped and spiny outside, the middle boat-shaped with an inflated concave membraneous margin, and the inner ring-shaped and wrinkled on the outside. HABITAT: fields, vineyards, waste ground; circum-Medit. Flowers nearly all the year round.

ECHINOPS—Globe Thistle
E. viscosus DC.
A pale blue flowered globe thistle growing to a height of 1½ m. with glandular hairy stems. Leaves twice cut into lance-shaped spiny segments, green above and grey-woolly below. Heads globular, about 5 cm. in diameter, made up of many 'headlets' which are 1-flowered and surrounded by involucral bracts; the outer bracts of each 'headlet' triangular, suddenly narrowed towards the base, the middle sharp-pointed and toothed, innermost bracts shorter, blunt and fringed. The leaves are very variable. HABITAT: fields and rocky places; Sicily, Greece to Palestine. June–August.

The similar *E. graecus Mill.* differs in having leaves much more finely divided into narrow segments, about 2 mm. wide, each ending in very sharp spines. The leaf segments are green and shining above with spiny teeth which spread out in all directions, like a gorse bush. Stems without glandular hairs. Flower heads amethyst-coloured with the outermost bract of each 'headlet' triangular egg-shaped with 5–7 teeth. HABITAT: dry, stony places and fallow fields; Greece. May–August.

E. ritro L. 205
GLOBE THISTLE

A very spiny plant with branched stems bearing large spherical spiny terminal heads of blue or whitish flowers. A perennial, up to 1 m. high, usually with a branched stem which is white and cottony but not glandular. Leaves shining green above and white-cottony below, leathery and deeply and irregularly cut into broad but very sharply pointed spiny lobes. Each 'headlet' surrounded by 20 blue bracts which are triangular-lance-shaped and hairless except for the hairy fringed margins. HABITAT: dry, rocky places and uncultivated ground; Spain to Palestine. July–September.

CARLINA—Carline Thistle
C. corymbosa L. 412
A stiff spiny biennial with thistle-like leaves and bright yellow flowers surrounded by spiny yellow bracts. Stems up to 40 cm. high, sparsely branched above with many stiff, wavy-edged, spiny-toothed leaves which are practically hairless, stalkless and half-clasping the stem. Heads few, sometimes solitary, 2½–4 cm. across; outer bracts leaf-like, dissected,

very spiny, green, spreading; inner lance-shaped, golden-yellow. HABITAT: garigue, uncultivated ground, clearing in thickets; circum-Medit. June–August.

C. lanata L.
Distinguished from all other Carline Thistles by the brightly coloured purple bracts surrounding the yellow florets. Flowering stem erect, 10–50 cm. high, branched or unbranched, with lance-shaped spiny cut leaves, the upper clasping the stem and all covered in close white woolly hairs. Flower heads also with woolly hairs, 1½–3 cm. across, often solitary; disk-florets yellow. HABITAT: garigue; rocky, dry, uncultivated places; circum-Medit. June–August.

ATRACTYLIS—Distaff Thistle

A. cancellata L.   413
An attractive small purple-flowered thistle with most unusual and delicate comb-like bracts enclosing the flower heads like a miniature lantern. A sparingly branched annual, 5–20 cm. high, with rather soft leaves covered in white cottony hairs. Leaves stalkless, narrow lance-shaped, uncut, bordered with spiny hairs. Bracts curved round and enclosing the flower head, green, and cut into long comb-like pointed spiny segments spreading on each side from a central midrib. Heads small, 1½ cm. long. HABITAT: dry, rocky places, track sides, hills; circum-Medit. April–July.

A. gummifera L.
has stalkless flower heads and compound thistle-like leaves forming a compact rosette pressed against the ground. Flower heads large, 3–7 cm. across, in the centre of the rosette of leaves and surrounded by spiny spreading bracts, the innermost tipped with red. Flowers purple. HABITAT: by roads and field verges; Corsica to Greece, Turkey, N. Africa. July–September.

JURINEA

J. anatolica Boiss. (J. mollis Asch.)
A rather soft woolly thistle with a stout white-woolly stem bearing a large terminal globose flower head, 3–5 cm. across, with rose-purple flowers. A perennial, 60 cm. high, with a rosette of basal leaves cut into lance-shaped widely separated segments, green above with cobweb-like hairs, white and softly hairy below. Stem leaves slightly cut or entire, with the blade often running for a short distance down the stem. Outer bracts of flower head prolonged into long green soft recurved teeth, and inner erect long-pointed with cobweb hairs. HABITAT: dry, grassy places, rocks and sunny hills; Dalmatia to Greece and Islands, Turkey. April–June.

CARDUUS

C. pycnocephalus L.
An annual thistle with rose-purple flowers bunched together at the ends of the stems and not over-topped by the leaves. Stems sparsely branched,

up to 1 m. high, with spiny lance-shaped toothed leaves with the blades running down the stem forming spiny wings. Leaves green above and white-woolly below. Flower heads cylindrical, 1–1½ cm. long, 2–5 at the tips of the branches; distinguished by the middle involucral bracts being lance-shaped, 2–3 mm. broad. HABITAT: cultivated ground, rough waste places, track sides; circum-Medit. April–July. Introduced into Britain.

The similar *C. argentatus L.* has flower heads usually solitary at the ends of the branches and the middle involucral bracts narrowly lance-shaped, 1–2 mm. broad. HABITAT: rough uncultivated ground; Greece to Palestine. April–June.

NOTOBASIS

N. syriaca (L.) Cass. **206**

One of the thistle-like plants common in the Mediterranean which have large rosettes of spiny leaves with striking white veins or white mottling. The flower heads are distinguished by the long and extremely spiny bracts, usually coloured silvery purple, surrounding the heads. An annual, 30–60 cm. high, with basal leaves oblong lance-shaped with shallowly-lobed wavy margins with many spiny teeth; stem leaves clasping, with 2 rounded wings, uppermost leaves under the flower heads cut into narrow sharp-pointed segments which radiate out beyond the flower heads. Flower heads egg-shaped, 2½ cm. long, with lance-shaped, long-pointed, purple bracts encircling the flowers. HABITAT: track sides, poor uncultivated ground and rough fields; circum-Medit. April–June.

SILYBUM

S. marianum (L.) Gäertn. **207**   HOLY THISTLE, MILK THISTLE

A biennial with a white-mottled or veined overwintering leaf rosette, a branched flowering stem 1–2 m. high, and few large purple flower heads surrounded by broad sharp-pointed bracts. Leaves shining dark green netted with white, deeply cut into triangular lobes with wavy, spiny-toothed margins, stalkless and encircling the stem. Flower heads 4–8 cm. across, purple, with hairless bracts broadening into egg-shaped spiny blades which taper into long stiff curved yellowish spines; inner bracts straight lance-shaped. HABITAT: rocky and uncultivated ground, track sides; circum-Medit. April–August. Naturalized in Britain.

The young leaves and midribs are a favourite salad plant among Arabs. Once eaten as a salad plant in England.

GALACTITES

G. tomentosa (L.) Moench **208**

A rather delicate thistle-like plant with narrow ovoid heads of attractive rose-purple or deep lilac flowers and narrow finely divided spiny leaves which are green and streaked with white above and white-hairy below. An annual or biennial 20–60 cm. high, branched above with lance-shaped leaves cut into lance-shaped spiny segments. Flower head 2 cm. by 1 cm. with the outer florets much larger than inner and spreading. Involucral bracts narrowed suddenly into a sharp fine erect spine which is grooved above; bracts with cottony hairs. HABITAT: dry places, uncultivated ground, sandy places; Spain to Greece, Crete, N. Africa. April–July.

CRUPINA

C. crupinastrum (Mor.) Vis. **414**

A soft spineless thistle-like plant with thin naked widely-spaced branches ending in solitary, rather narrow, cylindrical heads of a few purple florets. A hairless annual, 20–70 cm. high, with root leaves undivided but soon disappearing; lower stem leaves cut almost to the base into narrow linear widely spaced toothed segments which are thinly cottony underneath. Flower head rounded at the base, narrow egg-shaped, about 2 cm. long. Bracts of flower heads conspicuous, often purple-coloured, narrow lance-shaped, longitudinally ribbed with white papery edges, unequal in length and shorter than the flowers. Florets bunched together, not spreading, rosy purple. HABITAT: arid, uncultivated places, dry hills and amongst herbaceous vegetation; circum-Medit. (not Morocco). April–June.

LEUZEA

L. conifera DC. **209**

A conspicuous perennial with a large flower head like a pine-cone owing to the shining brown overlapping bracts, and with inconspicuous small purple florets. Stem 5–30 cm. high, covered in woolly white hairs, often with a solitary flower. Leaves green above and woolly white below, cut into narrow segments. Flower head ovoid, 3 cm. by 2½ cm.; involucral bracts prolonged into rounded papery cut extensions. HABITAT: garigue, rocks, pine woods and uncultivated ground; Spain to Italy, Sicily, Morocco to Tunisia. May–July.

CENTAUREA

C. pullata (L.) Cass. **212**

A biennial with large terminal pink flowers, conspicuous bristly-toothed involucral bracts and the narrow leaves close under the heads spreading beyond the petals. Very variable in height, single stemmed or sparsely branched, 1–30 cm. high, with stalked leaves cut into widely spaced triangular lobes. Flower heads 5–6 cm. across, often solitary, pale rose, purple, blue or white; bracts of involucre green with a narrow black margin and ending in a branched fish-bone-like spine of 4–10 bristles. HABITAT: fields, roadsides, screes; Spain to France, N. Africa; a casual farther east. April–June.

C. solstitialis L. **415**   ST BARNABY'S THISTLE

A white cottony-leaved thistle with rounded heads of yellow flowers and long yellow needle-like spines spreading outwards from the heads.

An annual with stiff cottony-winged stems, 20–60 cm. high, with basal leaves deeply cut into narrow toothed lobes, upper leaves narrow lance-shaped, spiny-tipped, with blades running down into wings on the stem. Flower heads 12 mm. across, solitary with involucral bracts spiny, with 1 long spine and several short basal spines. HABITAT: fields, cultivated places; European and Asian Medit., Egypt, Algeria. April–August. Introduced into Britain.

C. calcitrapa L.   STAR THISTLE
A thistle with rough but spineless leaves with reddish-purple flowers in an extremely spiny head covered in sharp spreading yellow spines. A biennial, 20–60 cm. high, with diverging branches arising just below the older flower heads. Lower leaves deeply cut into narrow lobes, upper irregularly toothed or entire, all with bristly pointed teeth. Flower heads 8–10 mm. across, old heads overtopped by later heads; bracts of heads with a long spreading spine, 2–2½ cm. long, and shorter spines at the base. HABITAT: track sides, dry fields and uncultivated ground; circum-Medit. May–July. Probably not native but introduced into Britain.

CARTHAMUS
C. arborescens L.   210
A giant thistle with large yellow flowers and very spiny bright green leaves and bracts. Stems up to 1½ m. high, with upper leaves ovate, half-clasping the stem, with very spiny irregular teeth and a long sharp-pointed apex; veins conspicuous, netted on the lower surface. Flower heads 5 cm. across with outer involucral bracts like the leaves and spreading to twice the width of the heads; inner bracts as long as the florets, spiny and similarly yellow coloured. HABITAT: sands and hills by the sea; Spain, Morocco, Algeria. April–August.

C. lanatus L. (*Centrophyllum lanatum DC.*)   211
An orange-flowered annual thistle with very spiny flower heads with cobweb-like hairs and spiny glandular-hairy and sticky leaves. An erect plant up to 60 cm. high, branched above, with dissected leathery spiny-toothed leaves, the upper clasping the stem. Flower heads large, globular, 2–3 cm. across, with very spiny lance-shaped dissected bracts, as long or longer than the flowers. Plant unpleasant-smelling, with reddish juice. HABITAT: rocks, dry hills, uncultivated ground and track sides; circum-Medit. May–August.

SCOLYMUS
S. hispanicus L.   213   SPANISH OYSTER PLANT
An extremely spiny, thistle-like plant with yellow stalkless flower heads surrounded and overtopped by very spiny bracts. A spreading biennial, 20–80 cm. high, with hairy stems and spiny-toothed wings running interruptedly down the stems. Leaves deeply cut, wavy margined and strongly spined. Flower heads in the axils of the leaves placed regularly along the

stems, 1½ cm. broad, and enveloped in 3 very spiny spreading bracts. Corolla without hairs; anthers yellow. HABITAT: waste places, uncultivated ground, sandy places; circum-Medit. May–August.

The similar *S. maculatus L.* has continuous wings on the stem, and wings, leaves and bracts all have thick white horny margins. Flower heads grouped at the ends of the branches; corolla with black hairs on the outside and dark anthers. HABITAT: rocky places, uncultivated ground; circum-Medit. May–August.

CICHORIUM—Chicory

C. intybus L.   CHICORY

A tall stiff non-spiny perennial with tough green stems and bright blue dandelion-like flowers ranged up the stems. Stem with widely spreading branches, 30–120 cm. high, with very variable leaves; basal short-stalked, lance-shaped and deeply cut or toothed, the upper clasping the stem, lance-shaped, distantly toothed or entire, all leaves roughly hairy below. Flower heads 2½–4 cm. across, in more or less stalkless clusters of 2–3 in the axils of the upper leaves; florets all ray-florets, a clear bright blue, rarely pink or white; pappus of scales $\frac{1}{10}-\frac{1}{6}$ as long as the fruit. A very variable species. HABITAT: track sides and uncultivated ground, usually on limestone; circum-Medit. May–September. Probably native of Britain.

The dried roots yield the chicory of commerce and are eaten as vegetables. Chicory is one of the 'bitter herbs' eaten with the paschal lamb. It has for a long time been regarded as an aphrodisiac; it is said to be valuable in pastures for the health of cattle.

C. pumilum Jacq. is similar but an annual with bluish-green leaves, a pappus ¼–½ as long as the fruit, and outer bracts of the flower heads densely glandular-hairy. HABITAT: rocks and sands by the sea; European and Asiatic Medit., Egypt, Algeria. May–September.

The closely-related endive, C. endivia L., is cultivated and often naturalized; its young leaves are eaten as salads.

C. spinosum L. is a very much branched shrubby perennial with spiny branches and only 5 florets in the flower head. HABITAT: a maritime plant of Greece, Crete and the Islands, Cyprus. June–August.

TOLPIS

T. barbata (L.) Gäertn.   **214**

A hawkweed-like annual with yellow flowers, often with a dark central 'eye' and unusual-looking thread-like bracts surrounding the flowers and spreading beyond the heads. A plant with widely divergent, practically leafless branches, up to 40 cm. high; leaves narrow lance-shaped, toothed. Heads of flowers yellow with the central florets often black, brown or reddish; stem below heads swollen and with long green thread-like bracts; involucral bracts green, soft, linear and drawn to a fine point. Outer fruits with scaly crown, inner with 4–5 hairs. HABITAT: dry places, uncultivated ground, sands; Spain to Greece, Crete, Morocco, Algeria. April–July.

RHAGADIOLUS

R. stellatus (*L.*) *Willd.* **416**

A small yellow-flowered hawkweed-like plant with very distinctive fruits which spread outwards into a 5-rayed star. A hairless or hairy annual, 10–40 cm. high, with few widely diverging branches and variable leaves, but usually lance-shaped and more or less toothed. Flower heads yellow, 1 cm. across, fruits rough, stiff, sharp-pointed, inner soon falling, outer persisting in a spreading star, about 2 cm. across. A very variable species. HABITAT: uncultivated ground, fields, rocks; circum-Medit. April–July.

HEDYPNOIS

H. rhagadioloides (*L.*) *Willd.* **417**

A small yellow-flowered hawkweed-like plant distinguished by its globular fruits and thick swollen stems below the fruits. A very variable plant, 10–40 cm. high, with sparsely branched, rough hairy leaves, lance-shaped, the lower toothed. Fruit a globular crown-like head formed by the curved hard hairy or hairless bracts which surround the fruits. HABITAT: fields and uncultivated ground; circum-Medit. April–July.

UROSPERMUM

U. picroides (*L.*) *Desf.*

A sowthistle-like plant with dandelion-sized bright yellow flowers and deeply cut leaves with irregular triangular lobes, the whole plant rough and bristly. An annual with sparsely branched stems, 30–45 cm. high, and upper leaves encircling the stem with pointed lobes. Stalk of flower heads naked; bracts of heads in one rank, fused below, egg-shaped, long-pointed and covered with long bristles; fruits curved, knobbly, with a long beak and terminal pappus. HABITAT: stony places, track sides, vineyards, uncultivated ground; circum-Medit.; March–June.

TRAGOPOGON

T. porrifolius L. (*T. sinuatus Aré-Lall.*) **215**   SALSIFY

A hairless, somewhat blue-green narrow-leaved biennial with dark reddish-purple flowers which open wide for only a short time during the day, and with green, pointed bracts usually spreading beyond the flowers. Stems erect, ½–1 m. high, simple or with few branches and with leaves narrow lance-shaped, parallel-veined, broadened at the base and half encircling the stem. Flower stalks strongly swollen under heads; involucral bracts 3–5 cm. long, often twice as long as the purple ray-florets. Fruits scaly and narrowed upwards into a long beak carrying the large parachute-like pappus and which collectively form a striking spherical 'dandelion-clock' when ripe. HABITAT: fields, meadows and dry places; circum-Medit. (not Turkey). April–June. Introduced into Britain.

The cylindrical tap roots are eaten as a vegetable, and various forms are cultivated for this purpose.

ANDRYALA

A. integrifolia L. **217**

A whitish-yellow softly hairy plant with close terminal clusters of lemon yellow flowers, with the stems, upper leaves and bracts covered in a dense pelt of woolly white or yellow hairs. An annual with erect stems, 30–60 cm. high, branched above, with narrow, softly hairy, lance-shaped and weakly toothed leaves, the lower stalked, the upper stalkless. Flower heads in somewhat flat-topped dense clusters; heads about 1 cm. across; involucral bracts narrow lance-shaped and covered in dense glandular woolly hairs. HABITAT: sands, rocks, track sides, heaths on siliceous soils; Spain to Italy, Morocco, Algeria. April–July.

LAUNAEA

L. anthoclada Maire **218**

A low shrublet with stiff zig-zag intertwined branches terminating in bright yellow heads of hawkbit-like flowers, about 2½ cm. across. Bushes up to 90 cm. high, with basal leaves very narrow lance-shaped with few widely spaced projecting teeth; stem leaves reduced to triangular papery scales. Flowers borne at tips of branches which later become spiny; involucral bracts in 2 ranks, the outer short egg-shaped, the inner oblong, 2–3 times as long, all smooth with white papery margins; pappus white. HABITAT: rocks and dry places by the sea and up in the hills; Spain, Morocco, Algeria. March–June.

CREPIS

C. rubra L. **216**   PINK HAWKSBEARD

Most species of this genus have yellow flowers, but this annual has large pink hawksbeard-like flowers. Stem 10–30 cm. high, simple or branched and few-headed with flower heads long-stalked, leafless and nodding before flowering. Leaves sparingly hairy, deeply toothed or lobed and narrowed into a long leaf stalk, upper narrow lance-shaped or linear. Flower heads up to 4 cm. across; involucral bracts 11–17 mm. long, outer sparsely woolly-haired, inner glandular-hairy. Fruits narrowed into a long beak. HABITAT: thickets among herbaceous plants, olive groves and rocks; France where naturalized, Dalmatia to Greece and the Islands; Turkey. April–June.

POTAMOGETONACEAE—Pondweed Family

Aquatic floating or submerged plants with usually opposite or alternate 2-ranked leaves with stipules. Flowers inconspicuous on small spikes with parts in 4's. Fruit usually 4 drupes, in our genus 1.

Posidonia oceanica (L.) Del.

An under-water marine flowering plant which often produces large numbers of fibrous balls called in France *pelotes de mer*. They are the fibrous bases of old leaves and they are washed up on the shore and often

195

form thick deposits along the drift line. It has a thick perennial stem covered in scales and reddish-brown fibres, recalling the tail of an animal. Leaves strap-shaped, dark translucent green, in 2 rows sheathing the stem, up to 50 cm. long by 5–7 mm. Flowers green, 2–4 in a group, and protected by a leafy bract of 2 valves; fruit fleshy, the size of an olive. HABITAT: growing to a depth of 30 m. in the sea; circum-Medit. October–June.

The leaves are sometimes collected and dried and used for stuffing mattresses and wrapping fragile objects. The leaves are also used as manure, being rich in salts and lime.

GRAMINEAE—Grass Family

One of the largest and most ubiquitous families, which dominates many parts of the world having low rainfall. Herbaceous annuals or perennials, rarely woody, with usually hollow stems with solid nodes. Leaves with a sheathing base and a narrow blade, and at the junction of blade and sheath there is often a flap of tissue or hairs called the *ligule*. Flowers clustered together in *spikelets* comprising generally 2 scale-like bracts (*glumes*) at the base, and 1 or more complete little flowers (*florets*) each consisting of an outer bract (*lemma*), an inner bract (*palea*), 3 stamens (rarely 2), and a 1-celled ovary usually with 2 feathery styles. Spikelets either stalked and arranged in loose clusters (*panicle*), or stalkless and clustered into a spike. The most important family of all to mankind for it produces the cereal crops, sugar, fermented beverages, fodder for cattle and stock, turf and building materials (bamboo).

BROMUS: spikelets many-flowered, more or less flattened, with often long-awned lemmas; glumes unequal, lower 3–5-nerved, upper 5–7-nerved.

AEGILOPS: spikelets compact, ovoid or cylindrical, of tough rounded glumes, each with 2–4 teeth or long points.

HORDEUM: spikelets in groups of 3 arranged broadside, alternately and closely on the central axis; glumes equal, 1-nerved, lemmas 5-nerved, awned.

ARUNDO: tall perennial reeds with leafy stems and large terminal plumes of pale soft densely hairy spikelets.

BRIZA: spikelets in spreading panicles on slender stalks, ovoid, swollen but somewhat compressed; glumes and lemmas blunt, broad, awnless, with rounded backs.

CYNOSURUS: flowering spike dense, rounded or cylindrical; spikelets of 2 kinds, the upper fertile, awned, the lower sterile of many pairs of stiff spreading spines.

POA: spikelets flattened, usually of 2–5 florets; glumes angled in section (not rounded-backed), awnless; lemma keeled, awnless, often with a tuft of hairs at the base and a papery tip.

VULPIA: flowering spikes very narrow, elongated, with short-stalked spikelets with long awns, arranged 1-sidedly; glumes unequal, lemmas tough.

CATAPODIUM: flowering spike simple or branched, rigid, with spikelets all on one side; glumes papery, 1–3-nerved; lemmas leathery, rounded on back, 5-nerved.

LOLIUM: flowering spike dense, oblong, flattened, spikelets arranged alternately edge-on to the kinked axis, with or without awns.

AVENA: stout annuals with large usually pendulous spikelets of 2–3 florets; lower glumes 7-nerved, upper 9-nerved; lemma with a long, jointed awn arising from its back.

LAGURUS: flowering spike ovoid, softly woolly; glumes with long spreading hairs; awn fine, kinked.

STIPA: flowering spikes usually long and wand-like; spikelets with very long awns which become spirally twisted below when dry.

CYNODON: perennials with flowering spikes of 3–7 spreading finger-like branches arising from the top of the stem, composed of many 1-flowered spikelets pressed closely to the axis.

DIGITARIA: flowering spikes of several spreading finger-like branches arranged near the top of the stem; spikelets arranged in 2 rows on each side of the flattened axis, usually in pairs.

ECHINOCHLOA: flowering spike branched with triangular-sectioned stems; spikelets plump, clustered along one side of the branches; sterile glume dry, papery, 5-veined, with or without an awn.

SETARIA: flowering spikes elongated cylindrical, bristly with 1 to several long bristles arising below and longer than the spikelet.

ANDROPOGON: perennial grasses with cylindrical flowering spikes in pairs, awned; spikelets also in pairs, 1 stalkless, fertile, the other male or sterile; awn kinked and twisted.

SORGHUM: robust grasses with large loose branched panicles of shining spikelets which are paired, 1 stalkless female and 1 stalked male; glume 3-pointed, becoming hard and shiny, lemma awned.

BROMUS

B. madritensis L. **418** (Fig. VIIIa)

A low-growing sparsely branched grass with very conspicuous large egg-shaped erect heads with many long fine awns giving the whole head a feathery appearance. An annual, 10–30 cm. high, with few flat soft leaves. Heads 4–15 cm. by 1½–6 cm. wide, usually rather dense, purple or green, of many short-stalked narrow erect spikelets each with 2 outer sterile glumes of very unequal length. Spikelets of 5–12 flowers with the lemmas ending in a fine awn up to 2½ cm. long. HABITAT: cultivated ground, dry waste places; circum-Medit. March–June.

AEGILOPS

A. ovata L. **419** (Fig. VIIIb)

At once distinguished by its rounded compact flowering heads of few spikelets with tough strongly ribbed outer glumes which usually end in 3 long very stiff awns. A low-growing tufted annual, 10–30 cm. high, with few short broad-based narrow triangular-bladed leaves which are much shorter than the leaf sheaths. Head narrow egg-shaped of 2–5 spikelets. Outer glumes thick-ribbed, commonly hairy and ending in 2–4 rough spiny awns up to 5 cm. long. HABITAT: in arid grassy places, and track sides; circum-Medit. April–June.

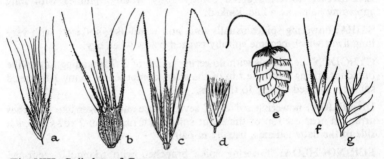

Fig. VIII. Spikelets of Grasses
a *Bromus madritensis* (×1) **b** *Aegilops ovata* (×1) **c** *Hordeum leporinum* (×1) **d** *Arundo donax* (×1) **e** *Briza maxima* (×1) **f** *Cynosurus echinatus fertile* (×1) **g** *C.e. sterile* (×2)

HORDEUM

H. leporinum Link **423** (Fig. VIIIc)

An annual with light green, weak leaves and dense cylindrical somewhat flattened barley-like heads with many long fine awns and bristles. Stems generally tufted, smooth, 6–60 cm. high, leaves rough, usually hairy and with slightly swollen sheaths. Flower heads 4–12 cm. by 1–3 cm., with spikelets in 3's arranged broadside to the axis. Spikelets with 2 long bristle-like sterile glumes, lance-shaped at the base with spreading hairs on the margins and long fine stiff awns; outer spikelets longer and wider than the median spikelet; lemmas all with long awns, 1–5 cm. long; groups of 3 spikelets usually falling together at maturity. HABITAT: track sides, waste places, dry uncultivated places; circum-Medit. April–June. Occasionally introduced into Britain.

The closely related *H. murinum L.*, the WALL BARLEY, which is a native British plant, is distinguished by the median spikelet being larger than the 2 lateral spikelets; in similar localities; circum-Medit. April–June.

198

ARUNDO

A. donax L. (Fig. VIIId)   GIANT REED, CANE

A tall bamboo-like reed with tough woody leafy stems thicker than a finger, and dense terminal plumes of pale flowers. Stems 4–5 m. high, commonly growing close together from underground tuberous stems. Leaves very long, 2–5 cm. broad, with smooth margins and ligule shortly hairy. Flower heads up to 70 cm. long, of numerous spikelets, about 12 mm. long, with hairless glumes and lemmas with long silky hairs. HABITAT: damp places and by streams; circum-Medit. August–December.

Probably originated in the Orient, but it has long been cultivated in the Mediterranean region where it is fully naturalized. Widely used for basket making, walking sticks, fishing rods, and for making windbreaks and shelters. It is the largest grass in Europe.

BRIZA—Quaking-Grass

B. maxima L. **420** (Fig. VIIIe)   LARGE QUAKING-GRASS

An attractive grass with neat plump egg-shaped pendulous spikelets on fine branches. A slender annual, 10–60 cm. high, with few stems and flat hairless leaves. Flower heads loose, sparingly branched, of few spikelets each 14–25 mm. by 5–15 mm. wide. Glumes broad, boat-shaped and neatly overlapping in 3–10 pairs, finely hairy, pale silvery green, often becoming brownish–purple. HABITAT: dry places and hills; circum-Medit. March–June. Introduced into Britain.

An ornamental plant both in the fresh and dried state.

CYNOSURUS

C. echinatus L. (Fig. VIIIf, g)   ROUGH DOG's-TAIL

Flower heads dense, egg-shaped, rather 1-sided with long shiny green or purplish awns. A hairless annual, 10–100 cm. high, with smooth stems and flat rough leaf blades. Spikelets densely crowded on stem, of 2 kinds, the outer sterile comb-like, of pairs of spreading narrow long-awned glumes, and partly concealing the fertile spikelets which have lemmas with fine straight rough awns, 6–16 mm. long. HABITAT: fields, rocks, grassy places; circum-Medit. April–June. Occurs occasionally in Britain.

POA

P. bulbosa L. **424** (Fig. IXa)   BULBOUS MEADOW-GRASS

A low-growing fine-leaved and tufted grass with moderately dense oval heads of tiny spikelets which are often replaced by miniature leafy shoots which drop off and root. Stems 5–40 cm. high, swelling at the base into 'bulbs'; these are often blown away in dry weather and can form new plants. Leaves very narrow, 1–2 mm. broad, usually folded and soon drying and breaking up. Flower heads 2–6 cm. by 1–1½ cm. wide, finely branched and bearing spikelets 3–5 mm. long, flattened elliptic and often becoming viviparous with long narrow leafy growths – *var. vivipara Koel.*

199

(**424**, Fig. IXa). Spikelets green, or variegated with white, purple and green. HABITAT: dry hills, rocks, track sides, largely on limestone; circum-Medit. March–June. A British native.

VULPIA

V. ciliata Link   (Fig. IXb)

A delicate grass with fine leaves and long narrow wand-like 1-sided heads of finely awned spikelets. Stems 10–40 cm. high, sheath of upper leaf encircling base of flower head. Spikelets with long awns, 1½ cm. long including awn, glumes unequal, the inner 2–6 times as long as the outer, awnless; lemmas with spreading bristly hairs. HABITAT: dry hills, sandy places; European and Asiatic Medit., N. Africa. April–June. Occurs occasionally in Britain as a casual.

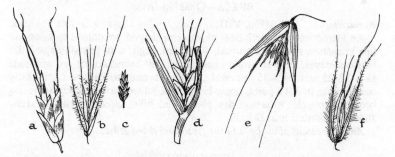

Fig. IX.   Spikelets of Grasses
a *Poa bulbosa* var. *vivipara* (×2)   **b** *Vulpia ciliata* (×2)   **c** *Catapodium rigidum* (×1½)   **d** *Lolium temulentum* (×1)   **e** *Avena sterilis* (× ½)   **f** *Lagurus ovatus* (×2)

CATAPODIUM

C. rigidum (L.) *C. E. Hubbard* (***Scleropoa rigida*** (L.) *Gris., **Poa rigida** L.*)
425   (Fig. IXc)   FERN GRASS

A short, tufted, rather ephemeral annual with fine leaves and erect stiff 1-sided heads of shortly stalked irregularly arranged spikelets. Stems many, 5–20 cm. high, flowering heads linear, or somewhat branched, spreading and egg-shaped. Spikelets mostly stalked, very small, 2–4 mm. long, linear-lance-shaped, several flowered; glumes with a fine tip, not awned; lemmas blunt. HABITAT: track sides, walls, dry places; circum-Medit. April–July. A British native.

LOLIUM

L. temulentum L.   (Fig. IXd)   DARNEL, BEARDED RYE-GRASS

A robust annual with erect narrow flowering heads with spikelets placed edge-on and arranged alternately along the stem. Stems solitary or tufted, 30–90 cm. high; leaves flat, 3–13 mm. wide, with a narrow auricle at the base of the blade. Spikelets about their own length apart, 1½–2½ cm. long, and appearing to be in the axil of a single broad rigid glume of about the same length. Lemma usually with a straight awn, but may be awnless. A variable species. HABITAT: among crops, fields; circum-Medit. April–June. A British casual.

Seeds sometimes contain a fungus and are then considered poisonous. Probably the tares of the Bible.

AVENA

A. sterilis L. 426 (Fig. IXe)   WILD OAT
A robust oat with loose heads of long-stalked, very large, conspicuous spikelets with broad pale glumes and long kinked awns. An annual, 60–150 cm. high, with broad flat leaves; flower heads becoming 1-sided after flowering. Spikelets 3–4 cm. long, outer glumes green with 7–11 veins, encircling the 3–4 fertile flowers. Lemmas covered with long silky hairs below, the 2 outer with long (4–5 cm.) awns with the lower part brown, stiff and twisted. HABITAT: fields, fallow ground, olive groves and vineyards; circum-Medit. April–June.

The closely related *A. ludoviciana Durieu* has smaller spikelets; it is naturalized in Britain and is spreading and becoming a serious pest.

LAGURUS

L. ovatus L. 427 (Fig. IXf)   HARE'S-TAIL
An attractive and striking grass with dense white silky flowering heads and fine projecting awns. A delicate, hairy, soft annual, 5–50 cm. high, with greyish-green leaves covered with woolly hairs and shorter than the rather swollen ribbed sheaths. Heads egg-shaped, generally about 2 cm. long by ½ cm. broad; spikelets narrow, 8–10 mm. long, 1-flowered; glumes hidden by dense white silky hairs. Awns very fine, 1–1½ cm. long, bent and twisted, purplish. HABITAT: dry places, particularly near the sea; circum-Medit. April–June. A casual in Britain.

Sometimes grown for ornament; the stems may be cut and dried for winter decoration.

STIPA

S. tortilis Desf. (Fig. Xa)
A slender annual with thread-like leaves and narrow contracted wand-like flowering heads and very long twisted awns. A rather tufted plant, 10–60 cm. high, with heads up to 12 cm. long, their bases encircled by the uppermost leaf sheath. Spikelets 2 cm. long; awns 8–10 cm. long, spirally twisted and bristly haired below, rough above, twice jointed. HABITAT: dry rocky places; circum-Medit. March–June.

CYNODON

C. dactylon (L.) Pers. 421 (Fig. Xb)   BERMUDA GRASS
A low greyish green creeping grass, often forming mats, with short erect

201

stems bearing a group of several narrow finger-like spikes. A perennial with stiff creeping stems covered in scaly leaves and rooting at the joints, bearing short leafy shoots which grow together in a mat. Leaves 2–15 cm. by 2–4 mm. wide with a row of short hairs where blade meets sheath. Stems slender, ending in 3–6 narrow purple or pale green spreading spikes each 3–4 cm. long. Spikelets 2–2½ mm. long, 1-flowered in 2 overlapping rows, egg-shaped and flattened. HABITAT: uncultivated ground, dry places, track sides; circum-Medit. May–August. Introduced to Britain a long time ago and more or less native in the West.

Widely used as a lawn grass in dry climates. Known as 'Kweek' in S. Africa, 'Doob' in India, 'Couch' in Australia and Bermuda Grass in U.S.A.

DIGITARIA

D. sanguinalis *(L.) Scop.*   (Fig. Xc)   HAIRY FINGER-GRASS
A rather weak soft-leaved annual grass with flower heads of several very narrow finger-like spreading branches arising close together at the top of the stem. Stems bent below and rooting at the nodes, 10–50 cm. high; leaves 4–9 mm. broad. Heads 4–10, very slender, 1–2 mm. across and 4–18 cm. long, often purplish. Spikelets in pairs arranged along one side of the branches, unequally stalked and 2½–3 mm. long, elliptic, awnless and finely hairy; lemmas 7-nerved. HABITAT: cultivated ground, damp sand; circum-Medit. June–November. Cosmopolitan, a casual in Britain.

ECHINOCHLOA

E. crus-galli *(L.) Beauv.* *(Panicum crus-galli L.)*   (Fig. Xd)
                                                        COCKSPUR GRASS
A luxuriant broad-leaved grass with irregular branched clusters of green or purplish, awned, usually bristly-haired spikelets. A tufted annual, 30–120 cm. high, with soft flat leaves, 8–20 mm. broad. Spikelets oval,

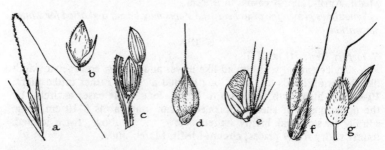

Fig. X. Spikelets of Grasses
a *Stipa tortilis* (×1½)   **b** *Cynodon dactylon* (×4)   **c** *Digitaria sanguinalis* (×3)   **d** *Echinochloa crus-galli* (×4)   **e** *Setaria glauca* (×3)   **f** *Andropogon hirtus* (×2)   **g** *Sorghum halepense* (×2)

3–4 mm. long, hairy, stalked, crowded in pairs or clusters, pointed or awned. Glumes unequal, the upper large 5-nerved with short spiny hairs on nerves; lemmas with a short point, or an awn up to 5 cm. long. HABITAT: sandy fields, track sides, rough places; circum-Medit. May–October. Cosmopolitan, a casual in Britain.

SETARIA

S. lutescens (*Weigel*) *Hubbard* (*S. glauca* (*L.*) *Beauv.*)   (Fig. Xe)

YELLOW BRISTLE-GRASS

An annual with terminal cylindrical heads of compact spikelets with many fine bristles spreading beyond the spikelets. Stems erect, 4–50 cm. high, solitary or tufted, leaves bluish-green, blades flat 3–7 mm. across, with finely pointed ligule or fringe of hairs. Flowering spikes 2–5 cm. long, with spikelets 3 mm. long, the lemmas of the fertile flowers with transverse wrinkles. Bristles arising from below the spikelets 5–10, rough, yellow or reddish-brown, up to 10 mm. long. HABITAT: dry places, cultivated ground; circum-Medit. May–September. A casual in Britain.

The similar *S. viridis* (*L.*) *Beauv.*, the GREEN BRISTLE-GRASS, is distinguished by its smooth lemmas and fewer greenish or purplish bristles; it has a similar distribution in the Medit. region.

ANDROPOGON

A. hirtus L. (*Cymbopogon hirtus* (*L.*) *Janch.*)   428   (Fig. Xf)

A tall tufted perennial with many stems and loose clusters of stalked silvery-hairy elongated heads borne in the axils of broad, partially enclosing sheath-like bracts. Stems ½–1 m. high, branched above, with flat grey-green leaves, 2–4 mm. wide. Flowering spikes paired, borne on hairy stalks 2–3 cm. long and arising from lance-shaped bracts. Spikelets 5–7 mm. long, covered with silvery hairs and with fine long kinked awns, 2–3 cm. long. HABITAT: in olive groves, dry hills and stony places; circum-Medit. April–August.

SORGHUM

S. halepensis (*L.*) *Pers.*   422   (Fig. Xg)

A robust broad-leaved grass with loosely branched erect heads of shining egg-shaped spikelets. A perennial with a creeping underground stem and erect stems, ½–2 m. high; leaves 1–2 cm. broad, with a prominent white mid-vein with a flat and hairless blade. Flowering head purple, loosely pyramidal, 20–25 cm. long; spikelets hairy, pointed or with a fine awn. HABITAT: sandy places, cultivated ground and rough places; European and Asiatic Medit., N. Africa. June–August. Introduced from the Orient but now widely naturalized.

MILLET, *Sorghum vulgare Pers.*, is sometimes cultivated in the Mediterranean.

203

PALMAE—Palm Family

A very abundant tropical family of graceful trees, shrubs and climbers with large pinnate or palm-shaped leaves often clustered into a crown at the end of a short or long stem. Flowers small, usually in 1-sexed clusters; the plants are often either male or female; clusters at first enclosed in a sheath; parts of flowers in 3's. Fruit mostly a berry or fleshy drupe. A family of great importance to people living in the tropics for it supplies food, clothing and housing, while copra, coconut and palm oil, dates, rattan, cane and raffia, etc. are of considerable economic importance. Many palms are grown for decoration around the Mediterranean.

PHOENIX: tall unbranched trees with large pinnate leaves and often edible fruits.

CHAMAEROPS: stemless suckering palms with flat fan-shaped leaves.

PHOENIX

P. canariensis Chaub.

This is the commonest ornamental palm which is widely planted along boulevards and coastal roads in the Mediterranean region. It is a native of the Canary Isles, but it grows vigorously in the Mediterranean climate. A tree with a stocky, thick stem, 6–8 m. high, covered in large closely-placed leaf scars. Leaves very large and widely spreading, arched outwards from a terminal crown, dark green, shining; leaflets narrow, sharp and stiff-pointed, very strongly keeled. Fruit small, brown, the size of an olive with coarse, fibrous, tasteless dry flesh. Introduced into the Riveria in 1864.

CHAMAEROPS

C. humilis L. **219**   Dwarf Fan Palm

A dwarf palm which is usually stemless and suckers from the base. Leaves fan-like with 12–15 lance-shaped, sharp, untoothed, spreading not drooping blades; leaf stalks slender usually with straight spines along the margins. Flowers in dense many-flowered heads among the leaves. HABITAT: dry, sandy and rocky places; Spain, Italy, Sicily, Sardinia, N. Africa. April–June.

A tough fibre is obtained from the leaves which is used for brush and rope making. The only native palm in Europe.

ARACEAE—Arum Family

Herbaceous plants with tuberous rhizomes. Flowers small, usually crowded on a club-like stem (*spadix*) and generally encircled by a leaf-like sheath (*spathe*), the whole forming the so-called 'flower'. Individual flowers usually 1-sexed, the males above, the females below. Ovary superior or sunk in the spadix with 1–many seeds. The often unpleasant smell of the flowers attracts pollinating insects.

204

ARUM: spathe large, funnel-shaped, constricted below the middle; spadix shorter, club-like; fruit a berry; leaves arrow-shaped.

DRACUNCULUS: leaves deeply cut into lance-shaped segments; spathe very large, constricted below the middle, with a spadix as long.

BIARUM: leaves long, narrow, entire, appearing after the flowers; spathe narrowed at the base into a cylindrical tube, blade oblong, acute.

ARISARUM: spathe fused into a tube for part of its length; spadix curved forward; leaves arrow-shaped, appearing with the flowers.

ARUM

A. italicum Mill. **220**

This lords-and-ladies is distinguished from our common British species (which also grows in the Mediterranean region) by the much yellower sheath and yellow club, and leaves which are usually spotted or veined with pale yellow. A perennial 20–60 cm. high with a large oblong tuber, and with long-stalked leaves; autumnal leaves are arrow-shaped with pointed lobes, the winter leaves larger, 15–30 cm. long, with rounded overlapping basal lobes, green-veined or spotted with pale yellowish white. Spathe enclosing flowers broad, up to 40 cm. long, golden yellow outside with inner surface whitish; spadix butter yellow, ⅓ the length of the spathe. Fruit scarlet, fleshy, in a spike. HABITAT: hedges, banks, woods and bushy places; Spain to Turkey; Morocco to Tunisia. April–May. A British native.

A. dioscoridis Sibth. **(224)** is somewhat similar but it has a very striking spathe, either purplish with deep brown-purple or blackish egg-shaped spots, or greenish yellow towards the apex and purple-spotted at the base. The spadix is blackish-violet. HABITAT: hedges, track sides; Turkey to Palestine. April–May.

A. palaestinum Boiss., which only grows in the Levant, has a conspicuous spathe, up to 18 cm. long, which is green outside and striking blackish-purple on the inside, and a blackish-purple spadix. An obnoxious-smelling plant. HABITAT: shady places amongst rocks and stone piles; Palestine. April.

DRACUNCULUS

D. vulgaris Schott (Arum dracunculus L.) **223**   DRAGON ARUM

A very striking sinister-looking plant with a stout stem mottled with dark purple blotches, a huge livid purple terminal 'flower', and stout deeply cut leaves. A herbaceous plant, up to 1 m. high, growing from a large globular tuber, with leaves with mottled sheaths encircling the stem, long stalks and deeply divided blades of finger-like lance-shaped segments, often mottled with white. Spathe very large, up to 35 cm. long, green on the outside but with a wavy purple margin, and deep chocolate-purple within; spadix thick, fleshy, often as long as spathe, similarly dark coloured. 'Flower' very fetid. HABITAT: woods, bushy places, uncultivated ground; Spain to Turkey, Algeria. April–June.

BIARUM

B. tenuifolium (L.) Schott **222**

A small arum-like plant with a brown-purple tongue-like 'flower' and a long narrow spadix of the same colour. Leaves with long sheaths, spatula-shaped and arising after the flowers. 'Flowers' practically stalkless with a spathe forming a pale cylindrical tube enclosing the individual flowers and prolonged into a long narrow lance-shaped blade 5–8 times as long as the tube; spadix narrow, cylindrical, often longer than the spathe. Female flowers separated by a gap from the upper male flowers, and above the latter are sterile flowers. HABITAT: rocky places; European Medit. (not France), Asia Minor, Morocco. April–July.

ARISARUM

A. vulgare Targ. Tozz. **221**   FRIAR'S COWL

An extraordinary-looking little plant with pulpit-shaped 'flower' striped with green and purple, and with a long curved club-like spadix projecting from it. A perennial with a short underground tuber and with long-stalked leaves arising from this. Leaves oval with rounded, heart-shaped, or pointed, arrow-shaped lobes; flower stem spotted and ending in a cylindrical spathe, with a flap-like projection curving over the opening, green-striped with greater or less amount of dull purple. Flowers 1-sexed, arranged at the base of a curved spadix which is greenish. HABITAT: hedges, vineyards and uncultivated places; circum-Medit. March–May and October–November.

LILIACEAE—Lily Family

Usually herbaceous plants, often with an underground bulb or corm, and often with basal leaves. Flowers often showy in few- or many-flowered heads. Parts of flowers commonly in 3's, usually with 2 similar whorls of petal-like segments, 6 stamens and a 3-celled superior ovary. Fruit a splitting capsule or berry. The family contains many ornamental plants, and a number of food plants such as onion and asparagus; its products include the drug aloin from Aloe species and the pesticide red squill from Urginea species.

COLCHICUM: petals united into a tube; styles separated from the base into 3 thread-like arms; flowers basal; rhizomes or bulbs.

ASPHODELUS: stems without leaves and all basal; petals all similar; flowers on single or branched stems; roots often tuberous.

ASPHODELINE: stems with many leaves; lowest petal narrower and a little distant from the others; stems many-flowered; rhizomatous.

ANTHERICUM: flowers white on both sides of the petals; petals spreading, with 3–5 veins; flowers on simple or branched stems.

APHYLLANTHES: flowers blue at the end of a rush-like stem without leaves.

GAGEA: flowers yellow, solitary or in clusters with leafy bracts below the flowers.

ALLIUM: plant smelling of garlic or onion; flowers in umbels at end of the stem with generally 2 encircling bracts.

FRITILLARIA: flowers large, solitary or few, or sometimes many in spikes; corolla bell-shaped, nodding and with a large glistening nectary at the base; flowering stems leafy.

TULIPA: flowers large, terminal, erect; corolla bell-shaped without nectaries; stems leafy at the base.

LLOYDIA: flowers tiny, white, solitary or several in a loose cluster; corolla funnel-shaped; stems leafy.

URGINEA: flowers many in a long terminal leafless spike; corolla spreading in flower; bulbs very large.

SCILLA: flowers blue, few or many in clusters or spikes; corolla bell-shaped or spreading in a star; petals with darker coloured mid-vein; leaves all basal.

ORNITHOGALUM: flowers usually white with a green line, in clusters or umbels; leaves all basal.

HYACINTHUS: flowers in a spike with leaves at the base; corolla funnel-shaped and fused together at the base to $\frac{1}{3}$ of their length; petals spreading and reflexed.

BELLEVALIA: flowers in a lax terminal head; corolla bell-shaped or tubular, not constricted into a throat and with 6 deeply divided teeth.

MUSCARI: corolla bell-shaped with 6 short teeth; flowers in dense clusters at the ends of leafless stems, upper flowers usually smaller, sterile.

ASPARAGUS: woody or herbaceous plants with leaves reduced to scales and branchlets needle-like, green, often in clusters; flowers 1-sexed with bell-shaped inconspicuous corolla.

RUSCUS: woody evergreen plants with stems flattened to leaf-like blades bearing flowers and fruits on one of their faces; fruit a red berry.

SMILAX: hooked climbing plants with heart-shaped leaves and tendrils from the leaf base; flowers in clusters; corolla small, bell-shaped.

COLCHICUM—Autumn Crocus

C. autumnale L. **230**   MEADOW SAFFRON, AUTUMN CROCUS

With pale purple crocus-like flowers which appear during the first rains of the autumn without any leaves. A perennial with a corm 3–5 cm. long with a blackish investing sheath, and with leaves produced in the spring in a group of 4 or more, 20–30 cm. long and 4–5 cm. broad. Flowers 10–25 cm. tall, solitary or in groups of 2–5, with very long stalk-like tubes; stigmas orange, each curved into a crosier or hook; stamens attached to petals at 2 different levels. The fruit is 3–5 cm. long and is carried above ground and encircled by leaves which soon wither. HABITAT: damp meadows; Spain to Greece, Algeria. August–October. Native in Britain.

C. neapolitanum Ten. is similar but rather smaller with the stigmas less curved and shallowly arched; stamens attached to petals at one level; flowers pinker with veins more or less wavy. Fruit enveloped by 3 leaves which are 13–20 cm. long and 1–2 cm. broad. HABITAT: dry sandy places; Spain to Italy, Algeria. September–October.

C. steveni Kunth is a small autumn crocus with clusters of 3–10 pink or white flowers on long stalk-like tubes which are 5–6 times as long as the petals. Leaves 5–7, very narrow, appearing at the same time as the flowers from a transparent enfolding sheath. Stamens yellow, shorter than styles. HABITAT: rocky places; Syria to Palestine. November.

C. cupani Guss. is another autumn-blooming species with rose-lilac flowers. Leaves 2–3, appearing with flowers, linear lance-shaped, 2–6 mm. broad, flat. Flowers 1–5, each 2–2½ cm. long with pointed petals and 5–11 veins; style scarcely longer than the dark anthers. HABITAT: rocks, grassy and stony places; Algeria, Tunisia, Italy, Greece and the Islands. September–November.

Colchicum species contain an important substance, colchicine, which if applied to seeds or growing points of plants has the effect of doubling the chromosome numbers of young dividing cells. It is of great importance in plant breeding for it causes permanent genetical changes in the tissues which may be of value in agriculture and horticulture. The autumn crocus was known to the ancients as a dangerous poison, but it has now gone out of favour as a medicinal plant.

ASPHODELUS—Asphodel

A. microcarpus Salzm. & Viv. (A. aestivus Brot., A. ramosus L.) 233

ASPHODEL

A large asphodel with a tall, much branched stem bearing a pyramidal head of many conspicuous white flowers with a reddish-brown vein to each petal. A hairless perennial with dahlia-like swollen tubers and a robust stem up to 1½ m. high. Leaves sword-shaped, up to 1 m. long and 2–4 cm. broad, V-shaped in section. Petals 1½–2 cm. long with the lower bracts longer than the flowers they subtend. Fruit the size of a large pea, elliptic, 5–8 mm. long. HABITAT: rocky places, hills and dry places; circum-Medit. March–June.

The presence of the asphodel is often a sign of derelict overgrazed ground; the plant is not usually eaten by animals. The tubers, rich in starch, are used to make glue for shoemakers and bookbinders. According to Theophrastus the root, stalk and seeds can be eaten; and Atchley records that during the First World War the roots were again eaten in Greece.

The similar *A. albus Mill.* (232) is distinguished by its narrow unbranched flowering stem (sometimes with a few short basal branches), and its compact simple spike of larger white flowers. Petals up to 3 cm. long, white with a brown vein; bracts subtending flowers lance-shaped, dark brown. Fruit larger, 1½ cm. long. HABITAT: dry places, meadows, thickets, hills; Spain to Greece. March–May.

A. fistulosus L. 238
A smaller, more delicate asphodel with narrow more or less hollow leaves and pink or white flowers with a reddish or green vein to each petal. A perennial with fibrous, not swollen roots, and a hollow stem, 20–60 cm. high, which is simple or sparsely branched at the summit. Leaves 1–3 mm. broad, linear, pointed, semi-circular in section with a narrow keel and roughish margins. Flowers about 2 cm. across in loose spikes with whitish bracts. Fruit globular, 4–6 mm. long, the size of a pea. HABITAT: fields, uncultivated ground, track sides; circum-Medit. March–June.

ASPHODELINE
A. lutea (L.) Rchb. 235   YELLOW ASPHODEL
An asphodel with a robust unbranched spike of large yellow flowers and many narrow leaves on the stem below. Stems up to 1 m. high with dense leaves along its whole length. Leaves stiff-pointed and triangular in section, sheathing at the base, smooth-margined, about 2 mm. broad. Flower head a dense spike, 10–15 cm. long in flower, and 30–50 cm. in fruit. Bracts at base of flowers papery, egg-shaped and pointed, conspicuous. Fruit 10–15 mm. long. HABITAT: rocky and stony places; Italy to Palestine, Algeria, Tunisia. April–May.

The similar *A. liburnica (Scop.) Rchb.* is a smaller and more delicate plant with very narrow leaves, about 1 mm. broad, and is easily distinguished by the fact that the upper third of the stem below the flower spike is without leaves. Leaves numerous on the lower part of the stem, with rough margins. Flower heads lax, long and narrow, with bracts much smaller and less conspicuous than in the preceding species. HABITAT: bushy and rocky places; Italy to Greece and the Islands, Cyprus. June–July.

ANTHERICUM
A. liliago L.   ST BERNARD'S LILY
An attractive lily-like plant with a single leafless stem bearing a few large pure white open flowers in a slender spike. A perennial, 30–60 cm. high, with leaves arising from the base, about as long as the flowering stem and 5–7 mm. broad. Bracts on flowering stem egg-shaped or lance-shaped, gradually narrowed at base and shorter than the flower stalk. Flowers 2–3 cm. long, petals oblong-lance-shaped, 3-veined, style curved upwards. HABITAT: in woods and dry grassy places, hills and rocks, on limestone; France to Turkey, N. Africa. April–June.

APHYLLANTHES
A. monspeliensis L. 252
An unusual-looking plant with a tuft of rush-like stems and small scaly

leaves, each stem terminating in a head of russet bracts bearing starry blue flowers. A perennial, 10–25 cm. high, tufted at the base with many thin somewhat grey-blue straight ribbed stems. Leaves reduced to a sheath encircling the stem. Flowers 1–3 in terminal heads, corolla blue or rarely white, with 6 spreading round-tipped petals each with a single vein of darker blue. Fruit with 3 valves enclosed in the papery bracts. HABITAT: arid and rocky places; Spain to Italy, N. Africa. April–July.

GAGEA

G. arvensis (Pers.) Roem. & Schult. 236

A tiny lily-like plant rarely more than 15 cm. high with a few flowers in a loose cluster which are yellow, and generally greenish on the outside of the petals. A bulbous plant with 2 tiny bulbs enclosed in brown papery scales. Basal leaves 2, channelled, with a rounded keel, 2 mm. broad. Flowering stem leafless below but with 2–3 leaves below the flowers. Flowers 1–1½ cm. long, 1–12 in an umbel-like cluster. The whole plant including the flower stalks is hairy. HABITAT: fields, vineyards, field verges and fallow; Spain to Palestine, N. Africa. February–May.

Of the numerous other species, which are difficult to distinguish from each other, *G. peduncularis (Pers.) Pascher* is distinguished by having the 2 leaves at the base of the flower head alternate and short-stalked, and leaves hairless or with hairs only on the leaf margin. HABITAT: hills and stony places; Greece to Palestine. February–May.

G. damascena Boiss. & Gaill. has 1 (not 2) bulbs enclosed in brown scales. Petals oblong-linear blunt, green with yellow margins. HABITAT: among rocks; Syria to Palestine. March–April.

ALLIUM—Garlic

A. ampeloprasum L. WILD LEEK, GREAT ROUND-HEADED GARLIC

A robust garlic with a stout flowering stem up to 1 m. high, leafy to the middle, and with a large globular head of many pink flowers on long stalks. A perennial with a strong garlic smell; bulb with many small bulbils. Leaves flattened, 12–35 mm. broad, very rough at the edges and on the keeled midrib beneath. Flower heads 7–10 cm. across, flowers rose-lilac with yellow anthers, and with 2 lateral projections from the stalk of the stamen, much longer than the anthers. HABITAT: hedges, banks and arid places; circum-Medit. May–July. Possibly native in Britain.

A. sphaerocephalum L. ROUND-HEADED LEEK

Flowers reddish-purple in a dense spherical head without bulbils, and hollow, cylindrical but grooved leaves. Flowering stem 30–80 cm. high, leafy on the lower half; leaves 1–1½ mm. broad. Flower heads 2–2½ cm. across, bracts below head shorter than the flowers. Flowers bell-shaped, about 5 mm. long; stamens protruding, the inner with 3 points with the central anther-bearing point rather longer than the lateral points. HABITAT:

dry places, cultivated and uncultivated ground; circum-Medit. June–August. A very rare British native.

A. subhirsutum L. 249

A rather delicate garlic with a rounded head of pure white spreading flowers borne on a cylindrical stem. Plant 20–50 cm. high, with a single stem bearing 2–3 soft flat broad leaves, ½–1 cm. across, with spreading hairs along the margins. Flowers 1 cm. across, white or with a pinkish stripe, stamens ⅓ shorter than the blunt petals; anthers pink; flower heads without bulbils. Bracts under flower heads 2–3 times shorter than the flower stalks. HABITAT: rocky, stony, arid places, woods; circum-Medit. (not Morocco). March–May.

A. trifoliatum Cyr. is very similar but the narrower, acute petals have a pinkish keel, or the flowers are pink throughout. The flower stalks are only 2–3 times longer than the flowers (in *A. subhirsutum* flower stalks 3–5 times longer), the stamens half as long as the petals, and the bract at the base of the flower head as long as the flower stalks. HABITAT: fields and hills; France to Palestine. April–May.

A. neapolitanum Cyr. 250   NAPLES GARLIC

A handsome plant with a spreading umbrella-shaped head of many pure white flowers on a triangular-sectioned stem. Flowering stem 30–60 cm. high, with 2 sharp angles and 1 blunt angle, bearing at the base 2–3 broad, flat leaves, 1–3 cm. across. Bract at base of head of flowers with 1 papery blade. Flowers long-stalked, 20–40 in an umbel; corolla cup-shaped with blunt rounded petals which are longer than the stamens. HABITAT: cultivated ground, olive orchards, fields; European and Asian Medit. to Egypt. Naturalized in Morocco. March–May.

A. triquetrum L. 248

A white-flowered garlic with a terminal head of rather few flowers which often hang to one side, and with a thick triangular-sectioned stem. Flowering stem with 3 sharp angles, 10–40 cm. high, bearing a head of 4–12 flowers with the bract at base of head 2-bladed. Flowers narrow bell-shaped, not opening wide, white with a green line, at first hanging to one side but later erect. HABITAT: cool shady places, by streams; Spain to Italy, N. Africa. February–May. Introduced into Britain and sometimes cultivated for its flowers which keep fresh after picking for several weeks.

A. roseum L. 251   ROSE GARLIC

A plant smelling strongly of garlic with beautiful, relatively large, pale rose-coloured or violet flowers in a rounded head. Plant 15–40 cm. high with a bulb surrounded by many small bulbils, and with the outer scales frequently perforated with small holes. Stem cylindrical with 2–6 flat linear leaves at the base, blade 4–10 mm. broad with fine teeth along the margins. Flower head rounded, with or without bulbils, with a 2–4-lobed bract shorter than the flower stalks. Petals narrow-elliptic, blunt, 10–12 mm.

long, much shorter than flower stalks; stamens shorter than the petals. HABITAT: grassy places, vineyards, stony hills, fields and on the littoral; Spain to Greece and Islands; Turkey; N. Africa. April–June. Introduced into Britain.

FRITILLARIA—Fritillary

F. graeca Boiss. & Sprun. 228

A small fritillary with solitary bell-shaped flowers which are dull brown-purple with a broad greenish band on the outside. Stem 30 cm. high, usually 1-flowered. Lower leaves rather broad, 8–20 mm. wide, the upper-most narrow lance-shaped or linear. Flowers 2½–3½ cm. long, petals blunt, outer oblong, inner a little broader and swollen at the tip; style with 3 stigmas. HABITAT: Stony and bushy places in the mountains; Greece. April.

F. messanensis Raf. 227

A fritillary with 1 or 2 rather large purplish flowers. Stem 20–50 cm. high, with 7–12 bluish narrow lanceolate leaves almost up to the flower, with sometimes a pair projecting above it. Flower bell-shaped, 2–4 cm. long, dark purple outside, sometimes with greenish bands, more or less lightened within by greenish yellow netting. A variable plant of confused nomenclature: the Spanish form illustrated – *var. hispanica* (*Boiss. & Reut.*) *Maire* – has rather egg-shaped flowers whereas the Greek form has longer, slightly reflexed petals. HABITAT: woods, scrub and meadows from sea level into the mountains; Spain, Italy, Greece, Morocco, Algeria. February–April.

F. acmopetala Boiss. (229) has globular-cylindrical flowers which are

somewhat narrowed at the throat, and the tips of the petals turn sharply outwards at right angles. Flowers nodding, usually solitary, 4 cm. long with outer petals greenish-yellow with brownish centres, strongly con-trasting in colour with the reddish-brown inner petals. Petals yellowish within and veined or chequered with red at the base. Plant up to 50 cm., leaves linear to lance-shaped, uppermost projecting well above flower. The colour of the flower is variable. HABITAT: field verges and stony places in the hills; Turkey to Palestine. April.

F. libanotica (*Boiss.*) *Bak.* 231

A tall species with leaves crowded at the base of the robust stem and a narrow spike-like head of many dull yellow or purple drooping flowers. Up to 1 m. high with whorled leaves, 2–3½ cm. broad. Flower spike 15–30 cm. long with 10–15 flowers; flowers 15 mm. long, greenish-purple or yellow-purple, drooping, on stalks longer than the flowers; petals blunt, obovate, often with darker veins and with a nectary at the base. HABITAT: dry stony places and walls; Turkey, Syria, Lebanon and Palestine. February–April.

TULIPA

T. clusiana Vent.   LADY TULIP

A beautiful tulip with the outside of the outer petals intense rose colour with white margins. Stems 20–40 cm. high, with 4–5 narrow lance-shaped channelled leaves. Flowers 3–6 cm. long, with ground colour white, inner petals shorter and more rounded, white with a violet base, outer petals longer pointed and brightly coloured. Stamens dark purple, without hairs. Bulb with dense woolly hairs on the inner side of the scales. HABITAT: often cultivated and naturalized in the Medit. region; a native of Syria and Persia. March–April.

T. praecox Ten.

A very striking species with bright scarlet flowers and a large black blotch at the base of each petal, with a narrow yellow band round each blotch. A robust tulip, 20–50 cm. high, with 3–5 oblong lance-shaped leaves with somewhat undulating margins. Flower bell-shaped, 5–9 cm. long, outer petals longer and acute, the inner with bluntly rounded tips; stamens dark purple, hairless. Bulb with dense felt-like hairs on the inside of the scales. HABITAT: cultivated and naturalized in many parts of the Medit., particularly Italy. March–April.

T. oculus-solis St Amans has similar flower colouring although it may be purplish red or yellowish on the outside, but it is distinguished by both inner and outer petals being acute, and the flower being less globular, and narrower and more funnel-shaped at the base, 3–8 cm. long. Plant 25–40 cm. high, with 3–4 broadly ovate-lance-shaped leaves. HABITAT: introduced to W. Medit. and now naturalized; native of Asia Minor to Palestine where it grows in fields. March.

T. boeotica Boiss. & Heldr. **225**

A handsome tulip with red flowers and bluish, wavy lanceolate leaves. The bell-shaped flowers on 20 cm. stems finally open wide and have a yellow zone at the base and an inner, star-shaped blackish purple blotch. Petals 5–6 cm. long, long-pointed, inner ones broader than outer. Stamens black. Bulb scales hairless. HABITAT: fields; Central Greece: common round Delphi. March–April.

T. sylvestris L. **226**

A yellow-flowered tulip with noticeably pointed petals. Stem 30–60 cm. high, usually with 3 narrow leaves. Flowers 3–5 cm. long with the petals of roughly equal length, the outer lance-shaped, the inner oval-lance-shaped without any basal spot, petals often greenish or reddish on the outside. Stalks of stamens yellow, hairy at the base, and longer than anthers. There are many sub-species and forms of this very variable species. HABITAT: cultivated ground, vineyards, meadows; Spain to Greece; Morocco to Tunisia. February–May. Naturalized in Britain.

LLOYDIA

L. graeca (*L.*) *Endl.*   **239**

A miniature lily not more than 15 cm. high and often much less, with grassy leaves and 2–6 white bells in a loose head. A bulbous plant with basal leaves linear, 2–3 mm. broad, and stem leaves much shorter and linear lance-shaped. Flowers stalked, at first nodding, and then erect; corolla funnel-shaped, 10–15 mm. long, white, with 3 purplish veins on the outside of each petal. HABITAT: dry rocky places; Sicily, Greece to Lebanon, Cyprus; Cyrenaica. March–May.

URGINEA

U. maritima (*L.*) *Bak.*   **429**   SEA SQUILL

An unmistakable plant either in the spring with its broad strap-shaped leaves growing out of huge bulbs, or in the late summer when the tall spikes of white flowers grow leafless from the dry ground. Bulbs up to 15 cm. across with broad lance-shaped, more or less flat shining leaves, 3–6 cm. across, appearing in the autumn and lasting till summer. Flowering stems robust, 1–1½ m. high, with very numerous stalked white flowers, in a long dense cylindrical spike; the lower flowers open first and flowering continues gradually upwards. Petals white, blunt and green-nerved; anthers greenish. HABITAT: sands, rocks and dry hills; circum-Medit. August–October.

The bulbs are sometimes collected for medicinal purposes; they are used in the treatment of heart disease, cough mixtures, etc., and have been known as medicinal since Pliny's time. The method of making 'vinegar of squills', a preparation in the British Pharmacopoeia, was described by Dioscorides. Red Squill rat poison is made from the bulbs with red tunics of certain North African sub-species. The Greeks hang the bulbs up in their houses during the New Year as a fertility rite. The bulbs unearthed by the plough are sometimes used to mark field boundaries. A good nectar-producing plant.

SCILLA—Squill

S. hyacinthoides *L.*   **244**

A robust plant up to 1 m. high with a long conical spike of many blue-violet flowers on long stalks of the same colour. Leaves numerous, very long, flat, rough, 2 cm. broad, with spreading hairs on the margins. Stem stout, leafless, bearing 50–100 flowers in a cylindrical conical spike. It has a large bulb up to 10 cm. long. HABITAT: hedges, rocky places and fields; European and Asiatic Medit. April–May.

S. peruviana *L.*   **243**

A handsome plant with a stout stem bearing a hemispherical head of very numerous violet-blue flowers, and with many broad strap-shaped basal leaves. Bulb very large; flower stems 20–50 cm. high and leaves longer than the stem, 4–6 cm. broad, and often with fine hairs on the margins. Flowers long-stalked, 50–100 in a head, sometimes white or pale

214

brown (N. Africa); stamens with swollen blue stalks. HABITAT: damp places; Spain to Italy, Sardinia, Sicily, Malta, N. Africa. April–June.

S. bifolia L. 240
A small squill with 2 shining channelled leaves and a lax head of 2–8 blue flowers. Flower stem 10–20 cm. high, with 2 (rarely 3) leaves half-way up the stem with bases encircling the stem. Leaves lance-shaped, blunt-tipped, 2–6 mm. broad. Flower head spreading, without bracts, with the lower flowers long-stalked; flowers and anthers blue. HABITAT: fields, woods and copses, ascending into the mountains; Spain to Palestine. March–June.

S. autumnalis L.   AUTUMNAL SQUILL
An autumn-flowering squill with thin stems and heads of small violet-purple flowers generally appearing before the leaves. Stem rough-hairy below, 10–25 cm. high, without bracts or leaves, bearing 4–20 flowers, at first grouped together and then spreading. Flowers 4–6 mm. long on spreading flower stalks. Leaves 5–6, very narrow, 1–2 mm. broad, slightly grooved and shining, appearing after the flowers. HABITAT: rocks and hills, dry grassy places; circum-Medit. August–October. A British native.

ORNITHOGALUM—Star of Bethlehem

O. arabicum L.
A rather robust pale yellow or dirty white flowered plant without the typical green band of the genus on the outside of the petals. Stems rather swollen, 30–60 cm. high, leaves very long, up to 60 cm., and 1–2 cm. broad, bluish green, without a white band. Flowers in a dense flat-topped head, flower stalks 4–7 cm. long and longer than the triangular-lance-shaped bracts. Petals 2–2½ cm. long, oval, blunt, with a fine point. HABITAT: fields, rocks and sands; European Medit., Chios, Palestine, N. Africa. April–May. Sometimes cultivated.

O. montanum Cyr. 237
A bulbous plant with hairless lance-shaped leaves and a short flattened spreading head of a few long-stalked white flowers with the outside of each petal with a central green band. A slender plant, 20 cm. high, with leaves 5–12 mm. broad, without a median white line and with small bulbs without bulbils. Flower heads of 3–20 flowers, petals 11–14 mm. long; fruit with narrow wings. HABITAT: grassy, rocky places; Sicily, Italy, Greece to Palestine. December–April.

The similar *O. tenuifolium Guss.* has very narrow leaves, 1–8 mm. broad, with a white median line. Bulb without bulbils; heads with 5–12 flowers; lower flower stalks 3–5 cm. long; petals 10–15 mm. long. HABITAT: rocks, garigue; France to Palestine, Morocco, Algeria, Cyrenaica. April–June.

The similar *O. umbellatum L.* is taller, 20–30 cm. high, with leaves 2–8 mm. broad, grooved, with white median line; bulb producing many bulbils. Flower head with 10–20 flowers; lowest flower stalks are 8 cm. long, and after flowering spread horizontally; petals 15–20 mm. long. HABITAT: cultivated ground, fields, stony places, groves; circum-Medit. April–June. Possibly native in Britain.

O. narbonense L.

A perennial with a long pyramidal many-flowered spike of rather small white flowers with petals with a narrow green band. Flowering stems standing above leaves, up to 60 cm. high; leaves spreading, channelled, broadly linear and somewhat bluish-grey, remaining green during flowering. Flowers 20–50, arranged regularly round the stem with bracts 1–2 cm. long, and much longer than the flower buds. Petals 10–12 mm. long; flower stalks at first spreading and then pressed against the stem. HABITAT: fields, grassy places; circum-Medit. April–June.

O. nutans L.   241

A perennial with a rather lax spike of large greenish flowers which hang in a one-sided head, and long soft green leaves. Flowering stems 30–60 cm. high, with long linear leaves, 8–15 mm. broad. Flower head of 3–12 drooping flowers, with stalks much shorter than the long pointed membraneous bracts; petals 2–3 cm. long, white within, and with a broad green band on the outside; stamens with stalk prolonged into 2 points between which lie the anthers. HABITAT: fields and vineyards; European Medit. and Asia Minor. April–May. Introduced into Britain.

HYACINTHUS—Hyacinth

H. orientalis L.   430

A very sweet-scented plant with a head of large drooping flowers, which is the parent plant of our cultivated hyacinths. Bulb large with pink, violet or white scales; leaves 4–5, broadly linear, dark green. Flowering stem thick, up to 30 cm. high, with 5–15 drooping flowers in a rather lax head. Flowers 2½ cm. long, blue, with short style and long anthers. HABITAT: cultivated places, fields, rocky places; native of E. Medit., Turkey to Palestine. February–April. Often cultivated and at times naturalized in other parts of the Mediterranean.

The 'Lily of the valleys', the 'lily that grows among thorns' of the Bible, probably refer to this plant. The plant we know as the lily-of-the-valley does not grow in Palestine, and the true lily, Lilium candidum, *is now rare and only grows in the Lebanon. Possibly a* Narcissus *or* Sternbergia *may be referred to, but the hyacinth is much the most probable plant.*

BELLEVALIA

B. ciliata (Cyr.) Nees (Hyacinthus ciliatus Cyr.)   242

A large bulbous plant with dull purplish yellow long-stalked flowers in a lax terminal head. The stout stem, 30–50 cm. tall, arises from 4 to 6 broad, glossy leaves 15–20 cm. long, 1½–2 cm. broad, densely ciliate on the margins. Flowers numerous, bell-shaped, pendulous, 8–10 mm. long, on spreading flower-stalks up to 8 cm. long which become rigid as the seeds form. HABITAT: fields, meadows; Italy to Palestine, N. Africa. Naturalized in S. France. March–April.

MUSCARI—Grape-Hyacinth

M. comosum (L.) Mill. (Leopoldia comosa (L.) Parl.) **245**

TASSEL HYACINTH

A large grape-hyacinth with a striking tuft of bright blue-violet sterile flowers at the apex and widely spaced russet-brown flowers below. Flowering stem up to 60 cm. high, a many-flowered elongated spike. Leaves broadly linear, 5–15 mm. across. Sterile flowers on long stalks curved upwards into a rounded or somewhat flat-topped head (in a closely related but very local Greek species, *M. pharmacusanum (Heldr.) Boiss.*, the sterile flowers are stalkless or very short stalked, and bright blue). Fertile flowers with short tubular bell-shaped corollas, 5–7 mm. long, on stalks as long as the flower or much longer; flowers erect and dull bluish purple in bud and turning greenish brown on horizontally spreading stalks. HABITAT: fields, cultivated ground, rocks, olive orchards; circum-Medit. April–June.

M. commutatum Guss. **247**

A dark blackish blue flowered grape-hyacinth with dense rounded heads of fragrant flowers with the lower flowers nodding, and with the teeth of the corolla 'pinched' inwards. Stem up to 20 cm. high, with a compact egg-shaped flower head, 2–4 cm. long, with a few (or no) sterile flowers at the apex, which are pale blue, small and shortly-stalked. Fertile flowers with a long egg-shaped corolla, 5–6 mm. long, which is 5-angled towards the mouth, and with short teeth with their points turned inwards and the same blackish-violet colour as the rest of the corolla. Leaves linear grooved, flaccid, longer than flowering stem. HABITAT: dry hills and grassy places; Sicily, Italy, Greece to Palestine. February–June.

M. atlanticum Boiss. & Reut. (M. racemosum auct.) **246**

A medium-sized grape-hyacinth which is easily distinguished by its dark blue flowers and contrasting dull-white teeth at the mouth of the corolla tube; sterile flowers paler blue, short-stalked and erect. Flower clusters dense egg-shaped, 3–4 cm. long, of 10–15 flowers each 4–5 mm. long. Leaves spreading over the ground, very narrow, 2–3 mm. across, semicylindrical in section and with a narrow groove. HABITAT: hills, fields and vineyards; circum-Medit. February–May. A British native.

Var. neglectum (Guss.) Brand. is distinguished by broader leaves, 4–5 mm. across, which are broadly grooved. Flowers 6 mm. long with a rather open throat. Fruit not flattened at the summit into a heart-shaped apex. HABITAT: common in fields, old walls and vineyards; circum-Medit. March–April.

217

M. pulchellum Heldr. & Sart. is very similar but a smaller plant with less constricted flowers with larger more conspicuous white teeth. Stem only up to 15 cm. high, with a rather laxer oblong flower head with 11–20 flowers; sterile flowers pale blue; lower fertile flowers short-stalked, later nodding, dark indigo blue with triangular white recurved teeth. Leaves very narrow, grooved. HABITAT: grassy hills and rocks; Yugoslavia to Lebanon. April.

ASPARAGUS

A. acutifolius L. 431

A woody-stemmed much branched scrambling plant, up to 1 m. high; leaves reduced to scales, and branchlets in clusters of 4–12, narrow, rigid and sharp-pointed, 5–10 mm. long and ½ mm. thick. Stems and twigs minutely hairy; lower scale-like leaves with a sharp spine. Flowers very small, green, bell-shaped; fruit black. HABITAT: dry places and hedges, largely on limestone; circum-Medit. August–December.

The very similar *A. aphyllus L.* has rough hairless non-climbing much branched angular stems. Branchlets spiny, in clusters of 2–6, ½–1½ mm. broad. HABITAT: stony sandy places and thickets; Spain, Sardinia, Italy, Sicily, Greece, Asiatic Medit., N. Africa. July-October.

The similar *A. stipularis Forsk. (A. horridus L.)* has solitary branchlets which are very stiff, spreading and sharp pointed, 2–3 cm. long and 1–1½ mm. broad, and these spines spread out in all directions from the stem. HABITAT: dry places, vineyards; Spain, Sardinia, Sicily, Greece, Cyprus, N. Africa. March–April.

RUSCUS

R. aculeatus L. 234   BUTCHER'S BROOM

A spiny evergreen bushy shrublet with tiny scaly leaves and branches flattened into leaf-like blades with small greenish flowers in the centre of the blades. Stems green, ribbed, much branched, 25–80 cm. high, with leaves small, brown, triangular and papery. Flattened branches thick, rigid, egg-shaped, 1–4 cm. long, and ending in a spiny point. Flowers 1-sexed, about 3 mm. across, 1–2 in the axils of papery bracts. Fruit a globular scarlet berry about 1 cm. across. HABITAT: thickets, dry banks and limestone; circum-Medit. September–May. A British native.

SMILAX

S. aspera L. 432

A climbing plant with hooked spines on the stems and broad shining leathery evergreen leaves, and heads of small greenish-yellow flowers and red berries. Stems angular, flexuous, sparsely spiny, climbing through shrubs and hedges to the height of 1 m. or more. Leaves alternate, heart-shaped, or with arrow-shaped lobes at the base and drawn out to a point at the apex; margin of blade and underside of midrib often with scattered spines; base of leaf-stalk with 2 tendrils. Flowers 1-sexed in small terminal

218

or axillary heads; fruit a red globular berry. HABITAT: hedges and thickets; circum-Medit. August–October.

The young shoots are eaten like asparagus.

AMARYLLIDACEAE—Daffodil Family

Bulbous plants with all leaves basal and flowers solitary or in an umbel, enclosed before flowering in papery bracts. Flowers with parts in 3's with 2 whorls of petal-like segments, sometimes with a crown-like projection (corona) in the throat of the corolla tube. Ovary inferior, 3-celled, with many ovules; fruit a splitting capsule. A family of many ornamental plants. Economically important in producing fibres like sisal, cuban hemp. Sugary exudates from Agave are distilled in S. America to make gin-like liquors.

LEUCOJUM: corolla bell-shaped, divided into 6 separate petals; flowering stem long, bearing one or a few flowers together, with 1–2 enclosing bracts.

STERNBERGIA: corolla in the form of a funnel and deeply divided into 6 erect petals; flowers 1 or a few, very short-stalked.

NARCISSUS: corolla with a cylindrical tube and spreading or reflexed petals, and with a cup-like or shallow ring-like projection (corona) at the throat; flowers solitary or several in a long-stalked umbel.

PANCRATIUM: corolla funnel-shaped and widening above with 6 long narrow petals and a corona with 12 teeth; flowers in a long-stemmed terminal umbel.

AGAVE: flowers numerous in a large branched spike from a rosette of narrow fleshy spiny leaves.

LEUCOJUM—Snowflake

***L. trichophyllum* Schousb.**

A pretty plant with 1–4 (usually 2 or 3) narrowly bell-shaped white or pinkish hanging flowers. Leaves 2 or 3, well developed at flowering time, grass-like. Flower stem 8–30 cm. high, usually longer than leaves, with 2 large membraneous bracts at top; from these appear the flower-stalks, which are unequal in length, erect and arching at the summit. Flowers usually 13–19 mm. long (up to 25 mm. in *var. grandiflorum (Red.) Willk.*), with narrow petals. HABITAT: open woods, sandy or gravelly pastures; S.W. Spain, Morocco. January–April.

***L. autumnale* L.**   AUTUMN SNOWFLAKE

A slender plant with 1–3 nodding white or pinkish open bell-shaped flowers, usually over before the grass-like leaves appear. Flower stem thin, up to 25 cm. high, with single small, membraneous bract. Flower-stalks unequal, erect, arching at the summit. Flowers 8–12 mm. long with oblong-elliptic segments. HABITAT: open woods, bushy places, dry pastures, marshes; Spain, Italy; Morocco to Tunisia. August–November.

219

S. lutea (L.) Ker. **253**

A beautiful crocus-like plant which flowers in the autumn and bears golden-yellow short-stalked flowers which appear at the same time as the leaves. Leaves strap-shaped, 5–18 mm. broad. Flowers 1, rarely 2, 3–5 cm. long with oblong-elliptic blunt petals on a very short tube, in the axil of a membraneous bract. Stamens ½ the length of the petals. HABITAT: fields, thickets and sunny hills; native of the Eastern Medit., but often escaped from cultivation and naturalized elsewhere. September–October.

The similar *S. sicula Tin.* **(254)** has narrower darker green leaves, with grey central band, 3–5 mm. broad, the margins glandular-hairy. Flowers smaller, nearly sessile, up to 3 cm. long, with pointed petals, and the stamens ⅓ as long as the petals. HABITAT: stony places; Sicily, Italy, Greece and the Cyclades. September–December.

NARCISSUS

N. tazetta L. **257**   POLYANTHUS NARCISSUS, ROSE OF SHARON

An attractive sweet-scented narcissus with a head of 2–13 flowers, with narrow milky white petals and a golden-yellow crown. Stems flattened, up to ½ m. high. Leaves linear, 5–15 mm. broad, keeled and bluish-green. Flowers with tube of corolla greenish, 1½–2 cm. long, petals obovate-elliptic, 12–25 mm. long, spreading or somewhat reflexed; crown 3–8 mm. long. A variable species. HABITAT: fields, meadows and garigue, especially in damp places; circum-Medit. November–April.

This narcissus grows abundantly on the plains of Sharon; bunches are brought into the houses, especially in Damascus. It may well be the 'rose' of Isaiah 35:1 – 'and the desert shall rejoice and blossom as the rose'.

The similar *N. aureus Lois.* has the tube of the corolla 2–2½ cm. long, yellow, with yellow petals and a deeper golden-yellow corona which is ⅓ as long as the petals. Stems up to 30 cm. long with 8–15 sweet-scented flowers; leaves 1–1½ cm. broad. HABITAT: field verges. France to Greece; N. Africa; January–March.

The similar *N. papyraceus Ker.*, the PAPER-WHITE NARCISSUS **(258)**, has the whole flower snowy white. Tube of corolla 1½–2 cm. long; petals egg-shaped, 12–15 mm. long; corona 4 mm. long. Stems with 5–20 sweet-scented flowers; leaves blue-green, broad, 7–15 mm. across. HABITAT: dry hills, vineyards, cultivated regions; Spain to Greece, Cyprus, N. Africa. January–May.

N. serotinus L.

A fragrant narcissus with white petals and a narrow golden-yellow crown, and very narrow rush-like leaves. Flowering stems slender, up to 25 cm. high, bearing a single flower (rarely 2), with a narrow green tube, 12–17 mm. long, short lance-shaped pointed petals, 9–12 mm. long, and a narrow 3-lobed crown, 1–2 mm. long. Leaves very narrow, 1 mm. across, usually absent during flowering. HABITAT: rocky places, dry hills and garigue; circum-Medit., except France. September–October.

PANCRATIUM

P. maritimum L. 433   SEA DAFFODIL

A handsome lily-like plant with broad bluish-green 'daffodil' leaves and a head of large white sweet-scented flowers, with the outer petals striped with green. It has a large bulb which sends up thick, broad (8–20 mm. across) blunt linear leaves, often spirally contorted. Flower stalk robust, compressed, up to 40 cm. high, bearing an umbel of 3–12 flowers, with 2 broad bracts subtending the flowers. Flowers large, up to 15 cm. long; tube of corolla 10–12 cm. long, expanding gradually into a funnel-shaped crown with 12 spreading teeth; petals narrow lance-shaped and spreading beyond the crown; petals fused at base to crown. Flowers rapidly followed by egg-sized green seed capsules. HABITAT: maritime sands; circum-Medit. July–October.

Theophrastus writes that the woolly hairs on the inside of the seed coat are used to weave felt shoes and other garments, and that the seeds are edible.

The similar *P. illyricum L.* has a flower with a narrower tube only 6–8 cm. long; broader leaves 3–5 cm. across, and heads of 6–14 flowers. HABITAT: shady rocks by the sea; Corsica, Sardinia and Capri. April–June.

AGAVE

A. americana L. 255   CENTURY PLANT

The enormous spiny rosettes of tough, excessively sharp-pointed leaves have now become a characteristic feature of the Mediterranean shore, and the great candelabra-like flower heads, which are produced once in the plant's lifetime, become temporary landmarks. Leaves of the rosette up to 2 m. long, rigid bluish-green and triangular in section, with tough spines along their edges. Flowering stems are produced after 10–15 years, and may grow to a height of 8–10 m. in a month. Flowers green with protruding yellow anthers, carried in many hundreds on the branched flower spikes. After flowering the plant dies but side shoots may grow on. HABITAT: naturalized on rocks, waste ground and the sides of roads, but commonly planted as a hedge round gardens and habitations from whence it has spread; now circum-Medit. June–August.

Original home Mexico, but it has been naturalized in the Mediterranean for over 2 centuries. In Mexico the national drink 'pulque' is made from the juice exuding from a severed young flower spike; it is fermented. The dry flower stems make good razor strops!

DIOSCOREACEAE—Yam Family

Usually slender herbaceous or woody-stemmed climbers with swollen tuberous roots. Leaves spirally arranged, often heart-shaped and entire, or lobed. Flowers small in 1-sexed clusters, male with 6 fused petals and stamens 6 or 3, female with inferior 3-celled ovary. Fruit a 3-valved capsule or berry. A large tropical family producing edible yams.

TAMUS

T. communis L.   BLACK BRYONY
A climbing plant with alternate heart-shaped long-pointed leaves, axillary clusters of small greenish flowers, and bright red shining berries. A perennial with a large blackish swollen rootstock which puts up annual twining stems, up to 3 m. high, with stalked leaves which are rather dark shining green, with 5–7 strong veins. Flowers 1-sexed, the male in longer spreading heads, the female shorter stalked and fewer flowered; corolla 3 mm. long, with 6 narrow somewhat recurved lobes. Berry 12 mm. across. HABITAT: woods, hedges; circum-Medit. February–April. A British native.

The rootstock has been used as a purgative and diuretic. The fruits have been used as a remedy for chilblains.

IRIDACEAE—Iris Family

Herbaceous plants with rhizomes, corms or bulbs and leaves which are often sword-shaped and sheathing the stem. Flowers usually large with parts in 3's. Petals usually fused at the base into a tube, stamens 3, ovary inferior 3-celled and style 3-lobed and sometimes petal-like. A family with many ornamental plants.

CROCUS: corm covered with usually fibrous bases of old leaves; leaves with a white midrib; flowers at first stalkless and ovary below ground; corolla a long slender tube with 6 petals.

ROMULEA: corm as in Crocus; flowers long-stalked on simple or branched stems; corolla with a short tube.

HERMODACTYLUS: similar to Iris but with tuberous roots; leaves tetragonal in section; flowers solitary and ovary 1-celled.

IRIS: plants with rhizomes or bulbs; flowers large, showy, with 2 whorls of petals, the outer 'falls' usually larger and reflexed, the inner 'standards' narrower and erect; the 'falls' are often 'bearded' with a tuft of hairs on the upper surface; styles broad, petal-like.

GLADIOLUS: flowers in a long spike, showy, inclined or horizontal with a very short curved corolla tube; bracts on flower stems usually leafy.

CROCUS

C. laevigatus Ch. & B.
Flowers white or lilac, feathered on the outside with violet lines, with the upper part of the throat golden-orange; stigmas feathery, orange. Corm with rather hard triangular scales covering the base. HABITAT: stony hills; Greece, Crete, Cyclades. November–January.

C. pallasii M.B. var. cartwrightianus (Herb.) Hay. 270

Flowers violet, 2–4 in a cluster, with the stalks of the stamens hardly longer than the anthers. Stigmas longer than stamens, orange. Petals blunt-tipped, twice the length of the tube, with inside of the upper part of the throat bright yellow, and with beard-like hairs. Corms covered with scales of a fibrous network. HABITAT: stony and grassy places; Greece, Crete, Cyclades. Autumn-winter, and spring-flowering in the mountains.

C. sativus L., the SAFFRON CROCUS, is closely related, and is considered by some to be a variety of C. pallasii. It is distinguished by its large orange-red stigmas which often project beyond the petals. Flowers lilac-purple, generally deeper purple at the throat. HABITAT: Possibly a native of E. Medit., and naturalized elsewhere. Autumn-flowering, with leaves appearing at the same time as the flowers.

The large stigmas produce the famous yellow saffron dye, and about 4,000 stigmas are required to produce an ounce of dye; used medicinally and for colouring cheese and other foods.

C. cancellatus Herb.

An autumn-flowering species with a coarsely meshed, tough fibrous network of scales covering the corm. Flowers white or pale lilac, often feathered with violet, with a hairless and often violet-streaked throat. Anthers yellow, 4 times as long as the stalk; stigmas divided into narrow, orange lobes, longer than the stamens. Leaves appearing after the flowers. HABITAT: rocky places in the hills; Greece to Palestine. September–November.

The corms are edible and are sold in the streets of Damascus and other cities.

C. chrysanthus Herb.

A deep yellow or orange-flowered species which flowers early in the year; the outer petals are often veined with purple. Corms with scales which detach in horizontal rings. Leaves appear with flowers; stigmas broadened at the end, scarcely divided into segments. HABITAT: grassy, rocky places; Greece and Asia Minor. January–February.

C. flavus Weston (C. aureus S. & S., C. moesiacus Ker.) 271

A bright orange spring-flowering crocus. Petals 38 mm. by 13 mm. broad, outer sometimes with greyish lines; stigmas shorter than anthers, orange; throat of flower without hairs. Corms rather large, about 19 mm. long, scales rather tough, later breaking into narrow linear fibres below. It is the parent species of the 'Dutch Yellow' commonly grown in gardens. HABITAT: stony slopes; Northern Greece and Asia Minor. January–February.

C. versicolor Ker.

A spring-flowering crocus with whitish or violet-grey flowers, richly marked with dark purple veins and sweet scented. Outer petals paler, whitish with violet veins, inner petals brighter coloured with darker veins;

throat white or pale yellow; anthers yellow and stigma orange. Corm with scales of parallel fibres, and sheath surrounding leaves and flowers often 2-lipped. HABITAT: garigue, hills and meadows; France and Italy. February–April.

ROMULEA

R. bulbocodium (L.) Seb. & Maur. **268**
A crocus-like plant with narrow rush-like leaves and funnel-shaped flowers. From a corm with leathery scales arises a short stem 2–3 cm. long, bearing 1–5 flowers, with a pair of ensheathing bracts below the flowers. Flowers longer than bracts, with throat and filaments hairy; petals lance-shaped, purple or lilac with orange throat, occasionally yellow or whitish. Stamens about ½ the length of the petals and shorter than the 3 stigmas which are each 2-lobed and divided to the base. HABITAT: dunes and sandy places by the sea; circum-Medit. February–April.
The similar *R. linaresii Parl.* has smaller violet flowers with purple throats. Sheaths below flowers distinctive; the outer is green and leafy, the inner transparent, papery, both little shorter than the flowers. Flowers about 15 mm. long with hairy throat; stigmas 3, shortly 2-lobed or unlobed, shorter than the stamens. HABITAT: maritime sands and turf; France to Turkey. March–April.

HERMODACTYLUS

H. tuberosus (L.) Mill. **262**   SNAKE'S HEAD IRIS, WIDOW IRIS
An unusual looking iris with bright yellowish-green flowers and dark purplish-black reflexed petals, and with narrow 4-ribbed rush-like leaves. A perennial arising from several spreading fleshy tubers, with leaves longer than the flowering stem which is 1-flowered and bears a broad leafy sheath up to 20 cm. long. Flowers fragrant, 4–5 cm. long, and with petals unbearded. HABITAT: bushy, rocky stony hills, garigue; European Medit. to Turkey, doubtfully native in Lebanon and Palestine, where it is frequently cultivated. March–April. Naturalized in Devon and Cornwall.

IRIS

I. pumila L. **260, 261**
A dwarf bearded iris with usually a single violet-purple or yellow flower on a short stem hidden by the rather narrow sword-shaped leaves, 7–17 mm. broad, shorter than the flowers. Stem creeping, cylindrical, thinner than a finger. Flowering stem, 5–15 cm. high, usually 1-flowered with 2 broad lance-shaped bracts with rounded backs and broad papery margins. Flowers with a tube 4–5 times longer than the ovary; petals 5–6 cm. long, the outer curved downwards, egg-shaped at the end, the inner oblong-ovate, erect. HABITAT: garigue, rocks, dry places; Yugoslavia–Greece. March–May.
Ssp. attica (Boiss. & Heldr.) Hay. **(260, 261)** has narrower leaves, 3–9

mm. broad, which are scythe-shaped. Flower a little smaller, pale yellow flushed with violet at the tips or intense violet, fragrant. HABITAT: rocky places in the hills; Greece. February–April.

The similar *I. chamaeiris Bertol.* (265, 266) has the tube of the flower less than twice as long as the ovary and hidden by the bracts; flowers violet or pale yellow, the inner erect petals nearly twice the width of the outer petals. *Var. italica Parlat.* has longer-stalked violet-purple flowers. HABITAT: garigue, rocky and dry places; France, Italy. March–May.

I. germanica L.
A very robust iris with a thick rhizome and broad sword-shaped leaves, with a branched stem, up to 1 m. high, bearing large blue-violet flowers. Leaves 1½–2½ cm. broad or more, shorter than the flower stem. Bracts below each group of flowers swollen, green at the base, dry and russet-coloured in the upper half. Flowers in groups of 2–3, 10 cm. across, with very broad petals, 4–5 cm. across, and nearly as long as broad; beard yellow. HABITAT: widely planted, particularly in Moslem cemeteries, but naturalized in many places; circum-Medit. April–June. There are many cultivated varieties.

I. florentina L. 259
A large white flowered sweet scented iris which is often flushed with pale blue. A robust plant, 40–60 cm. high, with branched stem bearing many flowers. Leaves shorter than flower spike, sword-shaped, 1½–2½ cm. broad. Flowers large; petals longer than broad, white with pale blue veins. Bracts below flowers boat-shaped, keeled and with dry papery margins, otherwise green. HABITAT: commonly cultivated but naturalized on rocks; probably of hybrid origin; now circum-Medit. March–April.

This plant is the model for the Fleur-de-lys of heraldry, and is the most important constituent of orris root which is used in perfumery.

I. cretica Jka. (*I. cretensis Jka.*) 264
An iris with very narrow grassy leaves and solitary blue-lilac flowers which are stalkless but with long pale tubes. Leaves all from the creeping rhizome, linear, 1–3 mm. broad, as long as flowers or longer. Bract below flower green, herbaceous, lance-shaped. Flower with a tube very much longer than the ovary; petals of equal length, blades ovate-oblong, blue-lilac, with a long narrow stalk which is yellowish with lilac veins; the inner petals are erect, the outer reflexed and unbearded. HABITAT: in bushy places, olive orchards, rocks; Greece to Turkey, Syria. February–April.

I. unguicularis Poir. (of which *I. cretica* is sometimes regarded as a subspecies) is the winter-flowering Algerian iris better known to gardeners as *I. stylosa Desf.* It is similar but more robust and with leaves 6–10 mm. broad. HABITAT: Similar to *I. cretica*; Algeria, Tunisia. December–January.

I. histrio Reichb. f. 267
A bulbous iris with beautiful solitary blue flowers with the outer turned-

down petals spotted with deep violet spots, and with an orange beard. Leaves 1–2, linear, tetragonal, appearing some time before the flowers and soon becoming longer. Tube of flower projecting beyond bract and 3–4 times longer than the ovary. Flowers 6–8 cm. across with outer petals oblong reflexed, and inner petals lance-shaped, broader towards tips, erect, blue and unspotted. Plant 12–20 cm. high. HABITAT: rocky places; Asian Medit., Turkey to Palestine. February–March.

I. palaestina (Bak.) Boiss.

A small iris with 2 files of outward-curving leaves, and yellow flowers tinged with pale purple. Plant with tuberous roots and very short or no stem, bearing 1–3 flowers. Leaves 8–12 mm. broad, folded longitudinally with horny and hairy margins. Bracts lance-shaped, long and tapering. Flowers 4 cm. long, the outer petals oblong with the upper quarter reflexed, inner petals small, narrow lance-shaped and spreading downwards. Stigmas prolonged into petal-like acute lobes which are divided into 2 points at the apex; flowers sweet-smelling. HABITAT: hillsides; Turkey to Palestine. January–March.

The similar *I. planifolia (Mill.) Fiori (I. alata Poir.)* is a much more conspicuous and larger flowered species with pale silvery mauve or lilac flowers and broad undulating turned-down petals with a conspicuous golden ridge down the centre of each. Stigmas erect, conspicuous, petal-like. HABITAT: stony and grassy places, thickets; Spain, Sicily, Sardinia; N. Africa, October–February.

I. sisyrinchium L. (Gynandriris sisyrinchium (L.) Parl.) 269 BARBARY NUT

A small iris with blue flowers with a white centre and rush-like leaves overtopping the flowers. Stem slender, wavy, up to 40 cm. high, arising from a rounded bulb covered in brown fibrous scales. Stem leaves 2, 3–8 mm. broad, linear and channelled, and rounded at back. Flowers in groups of 2–4 in the axils of dry, papery swollen bracts; corolla very variable in size, petals 2–3 cm. long; tube of flower longer than ovary. Outer petals spreading, broadly obovate, unbearded, bright blue, spotted with white or yellow towards the base, inner petals lance-shaped and erect; stamen and style fused in a column. Flowers usually opening only in afternoon. HABITAT: hills and dry places; circum-Medit. (not France). February–April.

I. xiphium L. 263   SPANISH IRIS

A bulbous iris with narrow channelled leaves and large solitary violet-purple flowers. Flowering stem 30–60 cm. high, longer than leaves which are awl-shaped. Flowers with a slightly swollen bract with unequal valves; tube of corolla very short and petals roughly equal in length, the outer a little longer than the stigmas and ending in a short oval limb; flowers easily broken off. HABITAT: fields by the sea; Spain, France, Corsica, Sardinia, Morocco, Algeria. April–June.

GLADIOLUS

G. segetum Ker.

A beautiful rosy purple flowered gladiolus with a number of flowers in a terminal spike at the end of a robust leafy stem. Stem 40–80 cm. high, arising from a globular corm covered in rather thick-meshed fibres. Stem bearing 3–5 broad leaves towards the base, and a terminal 1-sided spike of 6–10 flowers, each subtended by a bract which, in the lower flowers, is leafy and as long as the flowers. Flowers large, 4–5 cm. long, with noticeably unequal petals, the upper somewhat longer and nearly twice as broad as the lateral petals; anthers longer than their stalks. HABITAT: cornfields and cultivated ground; circum-Medit. April–June. Occasionally naturalized in Britain.

There are several closely related species growing in the Mediterranean region, e.g. *G. byzantinus Mill.* which has large rosy-purple flowers, 4–5 cm. long, with the petals more or less equal and contiguous and not separated as in *G. segetum*. Anthers the same length as the stalk; seeds compressed and winged. HABITAT: fields and uncultivated ground; Italy, Corsica, Sardinia, Sicily, N. Greece, Palestine, N. Africa. April–May. Occasionally naturalized in Britain.

G. communis L. (256) is very similar to *G. segetum* but the bracts are shorter than the flowers they subtend, and the petals are of almost equal size, the upper central one only slightly larger than the lateral petals; anthers shorter than their stalks. HABITAT: meadows and waste ground; Spain to Greece. April–June.

G. illyricus Koch has rather small purple flowers, 3½–4 cm. long, and bracts under the flowers shorter than flowers. Flowers with 3 upper petals unequal, the middle broadly oblong-elliptic and longer than the lateral oblong-ovate petals. Anthers shorter than stalk; stigmas narrow at the base and broadened suddenly into an oval blade. Leaves narrow, 5–9 mm. broad. Scales on corm with slender fibres. HABITAT: fields, bushy places and marshes; European Medit., Turkey, Lebanon, Palestine. April–June. A British native.

ORCHIDACEAE—Orchid Family

One of the vastest, oddest and most decorative of plant families. The Mediterranean representatives, unlike their exotic relations, are entirely terrestrial, mostly with tuberous roots, sometimes vertical rootstocks, or rhizomes.

The detail which follows refers specifically to the Mediterranean species. The leaves are entire, variously arranged, sometimes reduced to scales. The flowers are typically carried in spikes, often with bracts among them. The flowers are symmetrical if divided vertically, and consist of six segments in two sets, usually resembling petals though sometimes green. Of the outer three, referred to as *sepals* in the descriptions, one is vertical and the other two lateral. Of the inner three, two, referred to as *petals*, are placed between these sepals in the upper part of the flower. The remaining segment, the *lip*, is the lowest, and is usually much enlarged or elongated, often lobed in various ways. Some genera have a nectar-containing spur at the back of the lip.

The stamens and stigma of orchids are combined into a unique structure called the *column*. There are 1 or 2 stamens, which usually produce detachable pollen-masses or *pollinia*, and 2 or 3 stigmas; where there are only 2 fertile stigmas the third is converted into an organ called the *rostellum*, which is often beak-like in shape. The ovary is below the flower, and twisted, and the fruit is a dry capsule containing very numerous tiny seeds.

The study of orchids is made more confusing and more interesting by their erratic behaviour, some dying after flowering, others only flowering when conditions are suitable; by their readiness to hybridize; and in some cases, notably Ophrys, by their innate variability. Albinos occur in many genera.

The roots of some species are, or were, eaten by the local inhabitants: Lawrence Durrell records, for instance, the making of 'bulb tea' in Corfu from the tubers of Orchis laxiflora; the Greeks, French and Spaniards all use the Arabian word salep for orchids, which means the paste from orchid tubers ground up for food.

CEPHALANTHERA: with creeping rhizomes. Stem leafy. Flowers large, white or pink, with pointed segments. Lip with small serrated crests, spurless.

EPIPACTIS: with rhizomes and fleshy roots. Stem leafy. Flowers often inconspicuous, greenish or reddish, typically pendant, with triangular segments of similar size. Lip usually jointed in centre. Ovary not twisted, but flower-stems are. No spur. Late-flowering.

LIMODORUM: unmistakable, with very long, deep root, tall cane-like purple stem, leafless but with scales, and very large violet flowers with long spurs.

SPIRANTHES: with erect rootstock and fleshy roots. Stem leafy. Small white flowers in a spirally arranged spike.

GENNARIA: with a single tuber. Small plants with two heart-shaped leaves on stem. Flowers very small.

PLATANTHERA: with tapering tubers. Stem leafy, with two or three broad lower leaves; sheathing leaves on stems. Flowers large, whitish, fragrant, in loose spike. Lateral sepals outspread, upper sepal and petals forming an upright hood above the long narrow lip. Spur very long and slender.

NEOTINEA: with tubers. Small plants with few, usually spotted leaves. Flowers tiny, in a short, packed, one-sided spike. Segments forming a hood over 3-lobed lip. Spur very short.

OPHRYS: with rounded tubers. Leaves fairly broad, bright green, more or less pointed, ascending lower part of stem. Usually smallish plants with loose flower spikes. Flowers immediately recognizable by the large convex lip, often hairy in parts and frequently resembling an insect. Sepals outspread, green or pink, petals usually smaller; without spur.

Some species are constant in character but many are extremely variable, and individual specimens cannot always be assigned to a definite species; they also hybridize readily. This variation has given rise to very confused naming. In this book the classification and nomenclature suggested by Erich Nelson is followed. Leaf details are not given for individual species unless they differ from the norm; most of these plants are very similar in habit.

HIMANTOGLOSSUM: with tubers. Stem tall, thick and leafy. Flowers very large, in spike, with bracts. Segments forming a hood over the 3-lobed lip which is very large. Spur very short.

ORCHIS: with rounded tubers. Basal leaves in a rosette. Stem-leaves sheathing. Flowers usually in packed spike with membraneous bracts. Segments either all forming a 'hood' or 'helmet' over the column, or the lateral sepals only spreading outward or upward. Lip usually 3-lobed, with prominent spur.

DACTYLORCHIS: like *Orchis* but with lobed or divided tubers. Basal leaves not in a definite rosette at flowering time. Floral bracts leaf-like. Segments not forming a hood. Spur short.

SERAPIAS: with tubers. Stem leafy; leaves glaucous. Flowers unmistakable with a large pointed 3-lobed lip emerging at right angles below an elongated hood composed of the other segments, the sepals being joined together and the petals entirely concealed within them. Flowers in a spike, mingled with very large bracts. No spur. *Serapias* hybridize freely among themselves and also with *Orchis* species.

ACERAS: with tubers. Stem leafy. Flowers in a long narrow spike, with inconspicuous bracts, with segments forming a hood over a 3-lobed lip parted at the base and resembling a man. No spur.

ANACAMPTIS: with tubers. Stem leafy. Flowers small, in conical spikes with inconspicuous bracts. Segments outspread, lip 3-lobed, bearing two erect 'guide-plates' to direct the probosces of insects into the long thin spur.

CEPHALANTHERA—Helleborine

C. rubra (L.) Rich.   Red Helleborine

20–60 cm. tall, rather slender; with few, long, narrow, pointed leaves, widely spaced. Immediately recognizable by its beautiful, large carmine-pink flowers, 3 cm. across, with pointed, broad-based segments of almost equal size. Bracts very long. HABITAT: dry woods and clearings; Spain to

Greece, but near sea level only in Spain, France. April–July. Rare British native.

C. longifolia (L.) Fritsch (C. ensifolia Rich.) 434

SWORD-LEAVED HELLEBORINE

20–60 cm. tall, with long narrow leaves rising at an angle of 45° and in one plane on each side of the stem. The 10–20 white flowers, 20 mm. across, are carried erect in a loose spike. The sepals and petals all point forwards and inwards, forming an oval enclosure above the small lip, which carries an orange crest. Bracts much shorter than the flowers. HABITAT: shady places, in grass or under trees or scrub; circum-Medit. April–July. British native.

C. damasonium (Mill.) Druce (C. pallens Rich., C. alba (Crantz) Sim., C. latifolia Janch., C. grandiflora Gray) WHITE HELLEBORINE

20–60 cm. tall. Flowers like those of C. longifolia but fewer (3–12) and larger, not always opening properly, above oval leaves up the stem. Bracts longer than flowers. HABITAT: grassy places, woods, scrub, usually on limestone; circum-Medit., but only near coast in S. France; May–June. British native.

EPIPACTIS—Helleborine

E. helleborine (L.) Crantz (E. latifolia (L.) All.) 435

BROAD HELLEBORINE

25–80 cm. tall but often inconspicuous because of its overall greenish colouring. Lower leaves rounded, upper narrow, spirally arranged. Flowers 15–50 in a long more or less one-sided spike, drooping, widely opened, green, reddish or purple, 15 mm. across. HABITAT: dry woods, hedgerows, hillsides; circum-Medit. May–July. British native.

E. microphylla (Ehrh.) Sw. SMALL-LEAVED HELLEBORINE

20–40 cm. tall, very slender. Stem greyish green, pubescent near top. Leaves few, small, lanceolate. Flowers only 10 mm. across, rather bell-shaped, in a one-sided spike; greenish with purple tinges, slightly scented. HABITAT: stony places in woods and clearings; European Medit.; coastal mainly in S. France. May–June.

LIMODORUM

L. abortivum Sw. 288 LIMODORE

A very striking, unmistakable plant, forming a thick cane-like stem, 20–80 cm. tall, green entirely overlaid with violet, with leaves reduced to clasping scales. The large flowers, up to 40 mm. deep and wide, have wide sepals and long narrow petals, a large triangular lip, and a long down-pointing spur. Flowers violet with yellowish shadings; a yellow form is occasionally seen. The plant, which has a very deep, thick root, is by some considered a parasite on tree roots or may be saprophytic. HABITAT: rather local in dry open woods or clearings; circum-Medit. April–July.

SPIRANTHES

S. spiralis (*L.*) *Chev.* (*S. autumnalis Rich.*) 436 AUTUMN LADY'S TRESSES
A frail plant, seldom more than 10 cm. tall but can reach 30 cm. Immediately recognizable by its flowering time and the dense spiral of 4–5 mm. white flowers which are sweetly scented. The flower spike, which has leaf-like bracts, arises after the basal leaves die down, but a new rosette is usually formed alongside. HABITAT: dry grassy places, sometimes on sand dunes; circum-Medit. August–October. British native.

GENNARIA

G. diphylla Par. (*Gymnadenia diphylla Link, Coeloglossum diphyllum* (*Link*) *Fiori & Paol.*)
A stiff, insignificant plant 12–30 cm. tall, with only two heart-shaped leaves alternately on the stem. The numerous flowers, only 8–10 mm. long, are in a dense spike with small bracts. Flowers greenish yellow, hooded, with 3-lobed lip, the lobes pointed, and very short rounded spur. HABITAT: scrubby evergreen woods, shady grassy places; Spain, Sardinia; Morocco, Algeria. February–May.

PLATANTHERA—Butterfly Orchid

P. chlorantha (*Cust.*) *Reichb.* 437 GREATER BUTTERFLY ORCHID
A readily distinguished orchid, 20–50 cm. tall. Basal leaves usually 2, oval. The flowers, in a loose spike, are roughly cross-shaped, 18–23 mm. across, greenish white, pungently fragrant. Pollinia on column 3–4 mm. long, diverging. Spur very slender, up to 3 cm. long, curving down and forwards, slightly swollen at end. HABITAT: woods, scrub, open slopes, grassy places; circum-Medit. May–June. British native.

P. bifolia (*L.*) *Rich.* LESSER BUTTERFLY ORCHID
Very similar to *P. chlorantha* but smaller in all its parts, with narrower, closer-packed flower spike; the basal leaves very close together. Flowers 11–18 mm. across, more white than greenish, sweetly scented. Pollinia on column 2 mm. long, vertical and parallel. Spur not exceeding 2 cm., horizontal or down-curved, barely swollen. HABITAT: as for *P. chlorantha*; circum-Medit. May–July. British native.

NEOTINEA

N. intacta (*Link*) *Reichb. f.* 438
A slender, insignificant orchid, usually 10 cm. tall, rarely up to 40 cm., with few, usually spotted bluish green leaves, large in relation to the flower spike. Spike more or less one-sided, with small membraneous bracts; flowers numerous, palest pink or greenish white, only 6–8 mm. long, hooded, with forward-pointing, 3-lobed lip, the central lobe the longest.

Spur very short, conical, down-pointing. HABITAT: grassy places, hillsides, sometimes among scrub or trees; circum-Medit. March–May. Irish native.

OPHRYS

In the descriptions which follow the 'base' of the lip is visually its highest point, in the centre of the flower. The 'apex' or terminal lobe of the lip is its lowest point.

Note on habitats: All these 'insect orchids' grow in similar places, usually in grass or less often in stony places, in the open or among scrub, more seldom among trees, preferring calcareous soils.

O. fusca Link   274, 275, 276

With *O. lutea*, possibly the most widespread of the Mediterranean Ophrys. Clearly recognizable by its long, dark lip with blue 'eyes'. 10–25 cm. tall, with 3–6 flowers. Sepals green, broad, pointed, often much incurved; petals about ¾ length of sepals, yellow-green, strap-shaped. Lip elongated, with fairly pronounced, flattish side lobes and slightly indented terminal lobe; lip very deep brown approaching black, with two oval blue reflective eye-like patches near the base. Within each subspecies the variation is small.

Ssp. fusca Nels. (274) (including *O. funerea Viv.*), the most common, often has a narrow greenish or yellow margin to the lip; the latter is 11–15 mm. long. HABITAT: circum-Medit. February–May.

Ssp. iricolor (*Desf.*) *Schwarz* (275) is immediately recognizable by its much larger lip (up to 23 mm. long) with very large shining blue patch; the side lobes are marked, and the petals often reddish. Usually only 1 or 2 flowers on a stem. HABITAT: circum-Medit. February–April.

Ssp. omegaifera (*Fl.*) *Nels.* (276) is unique with the yellow W on the lip. The latter is broad and short (about 15 mm.), arched, with stubby, rounded side lobes, barely indented at the apex, medium brown and hairy below, the 'eye patches' coalescing, blue or brown, and bounded below by the W-shaped yellow band. Petals broad, often red-edged. Few-flowered. HABITAT: Crete, Rhodes, S. Spain, Morocco. December–March.

O. lutea Cav.   273

A very abundant species, easy to recognize by its lip which has a broad yellow margin to a central dark patch. 10–25 cm. tall, with 3–6 flowers, not varying greatly. Sepals green, broad, pointed. Petals ½ length of sepals, green, broadly strap-shaped. Lip oblong, with conspicuous side lobes and broad, indented terminal lobe, often flaring outwards. It has a wide, flat yellow margin surrounding an elongated dark or reddish-brown area carrying narrow-oval bluish reflective patches. HABITAT: virtually circum-Medit. with var. *minor* dominant from Sicily eastwards and var. *lutea* more common in the west, but there is considerable overlapping. March–April.

Var. lutea Nels. (273) is a much stouter plant with brighter yellow on the lip, which has a large basal lobe and broad plain yellow band, and is 12–18 mm. long.

Var. *minor Guss.* has a smaller terminal lobe with an inverted brown V at the base of the lip which is only 9–11 mm. long; sepals more rounded.

O. speculum Link 277   MIRROR ORCHID, MIRROR-OF-VENUS

Quite unmistakable with its blue-centred, hair-edged lip. Usually around 10 cm. tall, may reach 25 cm., with 2–6 flowers: very little variation. Lateral sepals narrow, green with maroon stripes, upper sepal hooded; petals very small, round, maroon. Lip 15–18 mm. long, pear-shaped, with arm-like side lobes (more prominent in the western than the eastern forms) and barely indented terminal lobe; the centre is entirely occupied by a blue reflective patch which has a narrow yellow border carried into the side lobes, the whole being edged with a thick fringe of dark red or blackish hairs. HABITAT: circum-Medit. March–April.

Var. *regis-ferdinandi coburgii Soó* is an even odder plant, similar but with a very narrow lip with a tuft of hairs at the base. The lip exactly resembles a bluebottle. HABITAT: Rhodes, probably Turkey. March–April.

O. cretica (Vierh.) Nels. 286

20–30 cm. tall with 3–8 flowers, not varying greatly. Sepals greenish, laterals dull pink on lower half; petals about ⅓ length of sepals, pointed, reddish. Lip about 16 mm. long, pear-shaped, with stubby, hairy side lobes, the central lobe with a prominent down-pointing protuberance. On a blackish-maroon background a long, more or less H-shaped white mark or pattern is outlined. HABITAT: Crete, Naxos, Karpathos. March–April.

O. argolica Fl. 272

10–25 cm. tall, with 4–8 flowers, not varying greatly. Sepals broad, pointed, pink with green central line; petals ⅔ to ¾ length of sepals, tending to be triangular, deep pink. Lip round or squarish, 15 mm. long; sometimes with flared edge, hairy at base, and small apical protuberance, sometimes with distinct hairy side lobes and pronounced protuberance. It is rich reddish-orange or brown with a small rectangular 'eye' pattern in white. HABITAT: Central Greece, Crete, Cyprus, Syria. March–April.

O. lunulata Parl.

30–40 cm. tall with 7–8 flowers, not varying greatly. Leaves rather long and narrow. Sepals pink with green lines, broadly triangular, pointed; petals almost as long, strap-shaped, pink. Lip 14–15 mm. long, narrow, much recurved, appearing almost tubular, with long hairy side lobes closely pressed to the central lobe which ends in a small forward-pointing protuberance. Lip dark red with a hint of yellow at the apex and a small sickle-shaped blue patch near the base. HABITAT: Sicily, Sardinia. March–April.

O. scolopax Cav. (including *O. oestrifera M. Bieb.*) 278, 284, 285
WOODCOCK ORCHID

Though this species is variable in its flowers there is a strong general resemblance in the lip, which is 9–15 mm. long, very rotund, ovoid to elliptic or oblong, with a forward-pointing protuberance at the apex. The side lobes are prominent, varying from round, hairy 'shoulders' to slender forward-pointing 'horns'. The lip is brownish or reddish brown with a complex pattern of lines and circles in white or yellow, sometimes enclosing a small blue patch. Sepals broad, rounded or pointed, tending to droop; petals about ⅔ length of sepals, more or less triangular and pointed. Plant 8–35 cm. tall. The most widely distributed ssp. are as follows.

Ssp. attica (*Boiss. & Orph.*) *Nels.* (285) has green sepals and petals. HABITAT: Greece to Palestine, Cyprus. March–April.

Ssp. scolopax Nels. (284) has sepals pale pink to purple-red, petals reddish. HABITAT: Spain, France, Yugoslavia, central Greece, Cyprus, North Africa. March–May.

Ssp. cornuta (*Stev.*) *E.G. Cam.* (278) has similar colouring to the last, and side lobes prolonged into forward-pointing horns up to 1 cm. long. HABITAT: Yugoslavia to N. Turkey. April–May.

O. fuciflora (*Crantz*) *Moench* (*O. arachnites* (*L.*) *Reich.*) 283

LATE SPIDER ORCHID

A very variable plant. Flowers with squarish lip with prominent humps. 15–40 cm. tall with rather stout, broad leaves and 5–8 flowers. Sepals broad, usually rounded, sometimes pointed, varying from white to pale or deep pink, with a median green line. Petals ⅛ to ¼ length of sepals, triangular, of similar colour. Lip squarish, often flared at the apex and sides, with a large greenish forward-pointing protuberance. Side lobes taking the form of prominent humps or points high up on the lip, which is around 15 mm. long, dark brown, rarely yellowish, with a variable yellowish pattern often enclosing a blue area in the upper part. *Forma maxima Fl.* (283) has very large flowers. HABITAT: Spain to Albania; N. Africa. March–May. Rare British native.

This description is of the type, *ssp. fuciflora Nels.* Three other subspecies have special characters and distinct distributions.

Ssp. candica Nels. has a broad, short lip 13 mm. long, a large whitish pattern enclosing a red-brown area and petals ⅕ length of sepals. Crete.

Ssp. oxyrrhyncos (*Tod.*) *Soó* has green sepals and tiny green petals; though the lip is normally like that of *ssp. fuciflora* it is on occasion pale yellow almost throughout, flaring into two rounded basal lobes with a very prominent protuberance. Sicily.

Ssp. exaltata (*Ten.*) *Nels.* (often regarded as a distinct species) is a bigger plant than the type; it has whitish pink sepals with three green lines, deep pink petals ½ length of sepals, and a narrower lip up to 20 mm. long with small side lobes and the pattern reduced to two small blue, yellow-edged patches or lines. Corsica, W. Italy.

O. bornmuelleri M. Schulze is a similar species with extremely short petals from Rhodes and the E. Medit. countries.

O. tenthredinifera Willd. **289**   SAWFLY ORCHID

Perhaps the most beautiful of the Ophrys, with its brilliant pink sepals and large yellow and brown lip, not varying greatly. 10–30 cm. tall, with 3–8 flowers. Sepals very broad, rounded, pale to rich carmine pink, with a green median line; petals ⅓ length of sepals, pink, triangular with rounded tips. Lip 16–20 mm. long, oblong in outline with side lobes reduced to rounded humps, the main lobe markedly flared towards the apex and divided into round lobes by a stubby protuberance. The lip is yellow or greenish yellow with a central red-brown area and a small white-margined blue pattern near the base, and is very hairy. HABITAT: Spain to W. Turkey; N. Africa. February–May.

O. apifera Huds. **290**   BEE ORCHID

15–45 cm. tall with 4–9 flowers, not varying greatly. This well-known plant has large rather blunt pink sepals with a green median line, and narrow green petals ⅓ to ⅔ length of sepals. The lip is 9–12 mm. long, oblong, with pronounced hairy side lobes in the form of forward-pointing shoulders, and a greenish, backward-pointing terminal protuberance. It is deep red-brown with a yellow pattern enclosing a pale red patch near the base, and often two yellow spots near the rounded apex. HABITAT: circum-Medit. April–May. British native. (*Sepals very pale in illustration.*)

O. bombyliflora Link **282**   BUMBLE BEE ORCHID

This small plant with squat-lipped flowers cannot be mistaken. It does not vary greatly. It is unusual in producing its new tuber on the end of a root several inches long. Sepals broad, usually rounded, green; petals ⅓ length of sepals, rounded, green. Lip squarish, 9–10 mm. long, broader than long; side lobes in the form of pronounced hairy humps which push out forwards and then project in a point behind. Lip uniform dull brown except for an ill-defined shiny, bluish area in the centre. HABITAT: probably circum-Medit. March–April.

O. sphegodes Mill. (sometimes spelt *O. sphecodes*) (*O. aranifera Huds.*)
                                 279, 280, 281   EARLY SPIDER ORCHID

15–60 cm. tall, with 3–10 flowers. A very variable species with an ovoid lip carrying more or less pronounced side lobes, varying from small humps to broad, angular projections near the apex or to 'arms' hanging beside the main lobe, sometimes hairy and sometimes smooth. The characteristic marking on the dark brown lip is a blue H pattern with the bar at the top, sometimes with a second bar below, sometimes degenerating into two blue lines, occasionally becoming roughly X-shaped. Petals up to ¾ length of sepals.

The group is segregated into numerous subspecies, of which *ssp. sphegodes* and *mammosa* are the most widespread, flowering February–April.

Ssp. **sphegodes** *Nels.* Sepals rather long and narrow, pointed, green; petals strap-shaped, sometimes waved, green or reddish. Lip lobes usually reduced to very small humps, marking often X-shaped; lip 10–12 mm. long. HABITAT: France, Italy, Corsica, N. Greece, possibly Morocco. Rare British native.

Ssp. mammosa (Desf.) Soó (**281**). Sepals rather long, rounded or pointed, often all green, laterals sometimes red below, rarely all pink; petals narrow, green or reddish, often waved. Lip lobes usually in form of pronounced forward-projecting humps, sometimes as 'arms'; lip round to oblong, 13–18 mm. long. HABITAT: Greece, Turkey, Crete, Cyprus, Palestine.

Ssp. atrata (Lindl.) E. Mayer. Sepals rather long, rounded or pointed, all green, petals broad at base, often waved, reddish green. Lip lobes usually projecting bosses; lip 12–15 mm. long. HABITAT: Spain to Yugoslavia.

Ssp. spruneri (Nym.) Nels. (**279**). Sepals rounded, drooping, laterals deep pink below, pale pink above, with median green line; petals almost as long, pale pink. Lip lobes in form of long drooping 'arms': lip 15 mm. long. HABITAT: S. Greece, Crete.

Ssp. litigiosa (Cam.) Bech. (**280**). Sepals drooping, pointed or rounded, green; petals broad, waved, green. Lip 9–11 mm. long; side lobes insignificant; main lobe edged with greenish yellow; blue marking usually rather small and neat. HABITAT: France, Yugoslavia, Corfu, Crete, Greece.

O. ferrum-equinum Desf. 287 HORSESHOE ORCHID

Usually recognizable by its blue horseshoe mark on the lip. 15–30 cm. tall, with 4–8 flowers, moderately variable. Sepals broad, blunt-pointed, clear or whitish pink with green median veins; petals narrowly triangular, pointed, almost as long as sepals, pink or carmine. Lip 14–16 mm. long, deep purple or purple-brown, typically with a horseshoe-shaped blue patch, points downwards, near the apex, sometimes reduced to two short vertical lines, rarely forming a square. HABITAT: Greece, Crete, Aegean Islands, Rhodes. March–May.

O. bertolonii Mor.

Readily distinguished by the blue patch near the apex of the lip. 15–35 cm. tall, with fairly long leaves, and 4–7 flowers, not very variable. Sepals blunt-ended, drooping, pink, often tinged green, with green median lines; petals ⅔ or ¾ length of sepals, narrowly triangular, pink. Lip almost black, 15 mm. long, sometimes up to 20 mm., oblong, bent forwards at the apex; side lobes sometimes arm-like, sometimes reduced to 'shoulders'; lip with a blue shield-shaped patch near the apex; latter indented and with a forward-pointing protuberance. HABITAT: S.E. Spain to Yugoslavia and Corsica. March–April.

O. arachnitiformis Gren. & Phil. FALSE SPIDER ORCHID

15–35 cm. tall, usually stout with very large leaves, and 3–6 or more flowers, extremely variable. Sepals broad, rounded, pink with green median line; petals fairly broad, rounded, sometimes waved, ½ to ¾ length of sepals, pink. Sepals and petals occasionally brilliant carmine. Lip 10–12 mm. long, squarish to oblong, with side lobes usually reduced to barely visible humps; main lobe rotund, with forward-pointing, apical protuberance. Lip rich reddish brown, sometimes tinged with yellow, carrying a variable pattern of white lines enclosing a blue patch, sometimes H or X shaped, sometimes tooth-like, sometimes a complex outline of partial circles. HABITAT: Spain to Italy, Sardinia, Sicily. March–April.

HIMANTOGLOSSUM

H. longibracteatum Schlecht. (Barlia longibracteata Parl., Loroglossum longibracteatum Moris.) **291**   GIANT ORCHID
A very massive orchid, with thick stem, 30–50 cm. tall, with large, tight-packed flowers 20–25 mm. long. Leaves large and shining, roughly oval. Flower-spike with narrow bracts longer than flowers, greenish or violet (like the upper part of the stem) according to lip colour. Upper segments short, loosely incurving, outer sepals erect like 'ears', reddish-violet. Lip greenish, reddish-violet or livid purple, three-lobed, the lateral lobes sickle-shaped, the wide central lobe divided into two short, pointed parts, all these divisions wavy-edged. Spur short, conical, down-pointing. Flowers usually but not always scented of lily-of-the-valley. HABITAT: usually in grass, sometimes in dry stony places; circum-Medit. February–April.

ORCHIS

O. papilionacea L. **292**   PINK BUTTERFLY ORCHID
The most handsome of Mediterranean Orchis species, 10–30 cm. tall. Leaves long, narrow, channelled. Flower spike loose, with 3–10 flowers up to 30 mm. long, among which are large pinkish bracts. Hood large, elongated, somewhat up-tilted. Lip large – sometimes as broad as long – fan-shaped, wavy-edged. Spur narrowly conical, shorter than ovary, usually down-pointing. Colouring of lines of deep pink or crimson on pink, red or violet ground, lip usually paler than hood. HABITAT: dry, sunny places, often among shrubs, or in grass; circum-Medit. March–May.

O. morio L.   GREEN-WINGED ORCHID
A variable plant, 10–40 cm. tall. Leaves oval-oblong to lanceolate, the lowest spread out or recurved, upper ones clasping the stem, which carries 6–20 flowers 15 mm. long, in a loose spike, becoming elongated. Bracts prominent, green or purple. Hood rounded over a lip widest at apex, divided into 3 shallow lobes, tending to be folded vertically. Spur short, blunt, slightly up-pointing, about as long as lip. Colouring very variable, from violet to pink or white, the lip spotted down the centre; lateral sepals generally veined with green.
Two well-marked subspecies, sometimes treated as species and largely replacing the type on the Mediterranean coasts: *picta (Lois.) Asch. &*

237

Graebn. is smaller in all its parts, lip not folded, colouring usually paler; *champagneuxii* (*Barn.*) *Cam.* always makes clumps, and has a sharply folded lip, whitish on the fold, lightly spotted; colouring pale purple or violet, spur long and white. HABITAT: dry and grassy places, woodland clearings; European Medit., Morocco, Algeria. March–June. British native.

O. longicornu *Poir.* (**O. longicornis** *Lam.*)   LONG-SPURRED ORCHID
Similar and closely related to *O. morio*, but more slender, only 18–25 cm. tall, fewer flowered. Flowers 12 mm. long. Lip white, purple-spotted in centre. Spur at least twice as long as lip, arching upwards. HABITAT: dry places; W. Medit. coasts and islands. February–April.

O. tridentata *Scop.* (including **O. commutata** *Tod.*) **296**
                                                    TOOTHED ORCHID
A very variable plant, 15–40 cm. tall. Leaves strap-shaped to lanceolate, glaucous, the upper ones clasping the stem. Flowers 10–15 mm. long, tightly packed into a rounded head. Bracts small, membraneous. Hood small, with the segments tapering into long points. Lip markedly 3-lobed, the side lobes like truncated arms, the central lobe shallowly notched or more often divided into two rounded, toothed sub-lobes with a central triangular tooth between them. Spur as long as ovary, cylindrical, down-pointing. Flowers fragrant, varying from red to pink, violet, lilac or white; hood streaked with purple lines; lip pink- or purple-spotted. HABITAT: grassy or scrubby places; circum-Medit. April–May.

O. lactea *Poir.* (**O. acuminata** *Desf.*) **297**
A variable orchid, related to *O. tridentata* and sometimes regarded as a sub-species of it. It is similar but differs as follows: height usually 10–30 cm., but plant robust; foliage sometimes spotted; segments of hood longer and narrower, tapering to diverging whiskery points; lip with side lobes often almost horizontal, central lobe fan-shaped, often undivid-ed; flowers often white or pale pink; lip often unspotted. HABITAT: dry grassy or scrubby places; probably circum-Medit. February–April.

O. purpurea *Huds.* **293**   LADY ORCHID
A robust plant, 20–40 cm. tall, rarely more, with large leaves varying from ovate-oblong to lanceolate, shining on top, paler on underside. Flowers 15–20 mm. long, in a dense conical spike which gradually elon-gates. Bracts reduced to purplish scales. Hood wide and forward-pointing forming a deep purple or wine-red 'bonnet' above the paler lip, which may be violet, pink or whitish, paler in the centre and spotted. Lip with two narrow arm-like upper lobes and a broad, skirt-shaped lower lobe, frilled at the base and cut into two lobes with a small tooth between. Spur short, curved downwards. The flowers vary a good deal in minor characters. HABITAT: dry, grassy places, woodlands; European Medit., Algeria, Cyrenaica. March–May. Rare British native.

O. italica Poir. (O. longicruris Link)   294

A very handsome orchid, 20–60 cm. tall, arising from a rosette of long, narrow, wavy-edged leaves, sometimes spotted. Bracts reduced to scales. Flowers 20 mm. long in a dense ovoid head, looking like little men with a narrow, forward-pointing hood, pink with purplish lines, above a long narrow lip, the upper lobes narrow, down-pointing, the main lobe with two strap-shaped terminal lobes, and between them a narrow-pointed tooth half as long. The lobes do not curl up as in *O. simia*, and are the same pale pink colour as the lip. Spur short and narrow, down-pointing. HABITAT: grassy or stony places in the open; circum-Medit. March–May.

O. simia Lam.   295   MONKEY ORCHID

20–50 cm. tall, with oval-oblong to lanceolate leaves. Flowers in a dense globular head, each 20–25 mm. long. The plant is well named, for each flower resembles a little monkey, the head represented by a rounded hood, the arms and legs by long curving strap-shaped lobes. The lobes are usually dark red in contrast to the whitish lip centre and silvery pink 'helmet', though the colouring varies. The two lower lobes are separated by a short, pointed tooth. Spur short and swollen, down-pointing. HABITAT: dry, grassy places, scrub; circum-Medit. March–June. Rare British native.

O. coriophora L.   439   BUG ORCHID

A slender orchid 20–40 cm. tall with long narrow-pointed leaves and a dense flower spike, with an unpleasant smell like that of a bed bug. Bracts 1 cm. long, decreasing up spike. The flowers, 10–12 mm. long, have a pointed hood, brownish purple with green streaks, above the wine-purple lip, which is paler and often greener near the centre. The lip is 3-lobed, the central lobe much longer than the side ones. Spur short, arched downwards. *Ssp. fragrans (Poll.) L.* is smaller, with central lip lobe more pronounced, and with an agreeable vanilla odour. HABITAT: grassy places, dry or damp; circum-Medit. April–June.

O. sancta L.   HOLY ORCHID

A very curious plant 15–45 cm. tall, in habit like *O. coriophora* and by some authors considered a subspecies of it, but quite distinctive with flowers twice as large – up to 20 mm. long – of a pinkish red, and with the lateral lip lobes each divided into up to 4 separate teeth. HABITAT: sandy and grassy places; Aegean Islands, Turkey, Syria, Lebanon, Palestine. April.

O. collina Soland. (O. saccata Ten., O. sparsiflora Ten.)   298

Robust but only 10–20 cm. tall, with broad, deep green leaves, often spotted, upper leaves clasping. Spike with 3–18 rather large flowers, 15–20 mm. long. Bracts small. Hood composed of dorsal sepal and petals only, the lateral sepals erect or projecting backwards. Lip entire, obovate or rounded, sometimes fan-shaped, wavy-edged. Flowers purplish-red,

lip paler than the rest. Spur white, short, conical, down-pointing. HABITAT:
dry grassy places; circum-Medit. January–April.

O. laxiflora Lam. 299   LOOSE-FLOWERED ORCHID
An impressive plant, 30–60 cm. tall, occasionally up to 100 cm., with
long, narrow, keeled, erect leaves, shining on top, bluish below, and a
long loose spike of 6–20 flowers 15–20 mm. long, backed by red or purple
bracts. Lateral sepals held outwards or backwards, dorsal sepal arched
over forward-pointing petals, above a roundish lip which is typically
folded at the sides, wavy-edged, slightly 3-lobed, the centre lobe where
present shorter than the outer ones. Spur long, cylindrical, blunt-ended.
Flowers typically claret-red or violet-red, sometimes pink or white.
HABITAT: marshes, damp meadows; circum-Medit. March–June. Britain:
Channel Islands only.

O. palustris Jacq.
Very similar to *O. laxiflora* and often growing with it. The spike is
narrower; the lip is more markedly 3-lobed, the central lobe longer than the
side ones, and the lip is not folded back until the flower is about to fade.
The colour is nearer magenta, with a white centre carrying small purplish
markings. HABITAT: marshes, damp meadows; circum-Medit. April–June.

O. mascula ssp. olbiensis (*Reut.*) *L.* 300
A slender Mediterranean form of the Early Purple Orchid, 10–25 cm. tall,
with oblong-lanceolate leaves, sometimes purple-spotted. Bracts small.
Flowers 5–15, 15–20 mm. long, in a loose spike; lateral sepals erect, other
segments forming a hood over a prominently 3-lobed lip which is folded
back at the sides. Lip whitish, purple-spotted in centre. Flowers otherwise
pink or violet-pink. Spur about as long as ovary, cylindrical, up-pointing.
Lacks the tom-cat odour of the Early Purple. HABITAT: grassy hillsides;
Spain to France, Corsica, Majorca; Morocco, Algeria. March–May. Type
is British native.

O. anatolica Boiss. 301   ANATOLIAN ORCHID
A slender plant not more than 25 cm. tall, with oblong to lanceolate leaves
and a loose spike of 5–8 flowers 12 mm. long, of pale rosy purple,
distinguished by the long more or less 3-lobed lip and the very long
cylindrical, pointed spur, horizontal or up-pointing, much longer than the
ovary. Bracts shorter than ovary. HABITAT: grassy or stony places; Turkey
to Palestine, Cyclades, Aegean Islands. Reported from Tunisia. March–
April.

O. provincialis Balb.   PROVENCE ORCHID
One of the few yellow-flowered orchids, usually 10–20 cm. but sometimes
up to 40 cm. tall, with narrow, pointed, glaucous leaves, usually purple-
spotted, and a loose spike of up to 14 flowers 18–20 mm. long, with
lateral sepals erect or projecting backwards, the dorsal sepal erect. The
slightly convex lip has three pronounced lobes; its edge is often frilled.

240

Spur almost as long as ovary, thick, cylindrical, swollen at end, curved and up-pointing. Flowers very pale yellow or cream, rarely greenish or pinkish; the lip usually lightly purple-spotted. There is a faint odour of elder.

Ssp. *pauciflora* (*Ten.*) *Balb.* (305) is a more robust plant 10–20 cm. tall, with leaves almost always unspotted. Flowers up to 8; lip brighter yellow, orange in centre, with small brown spots. HABITAT: woods, shady grassy slopes; circum-Medit. April–June.

O. quadripunctata Cyr. 302   FOUR-SPOTTED ORCHID

A very slender, graceful plant 15–25 cm. tall with small narrow leaves and a loose, elongated spike of 8–25 flowers 8–10 mm. long. The lateral sepals are erect; the rather rectangular lip is 3-lobed. Spur very long and thin, down-pointing. Flowers pink flushing to violet at the edges of the lip, which is whitish in the centre, with 2 or 4 small deep purple spots. HABITAT: stony hillsides; Italy to Turkey, Sardinia, Cyprus. April–May.

DACTYLORCHIS

D. romana Seb. (*Orchis sulphurea Link*)   303, 304

A handsome plant 15–35 cm. tall with long narrow leaves and a loose spike with prominent erect bracts at least as long as the flowers (15–20 mm.). Sepals erect, petals pointing forward and coming together. Lip rounded to oblong, with three lobes, the central one larger, often wavy-edged. Spur at least as long as ovary, usually pointing up almost vertically. Flower varying from violet or carmine to yellow or white. HABITAT: rocky or sandy places, even screes; circum-Medit. excluding France. March–May.

SERAPIAS

S. cordigera L. 308

A sturdy plant, 15–45 cm. tall, with narrow, channelled, pointed leaves, arching inwards; the stem reddish at the top, spotted at the base as are frequently the lower parts of the leaves. This distinguishes it from the other species immediately. Flowers 25–40 mm. long, 3–10 in a short spike. Bracts shorter than flowers, of similar colouring, or sometimes grey. Hood reddish violet or wine coloured, paler outside. Lip blackish purple, hairy, the central lobe very large and pointed heart-shaped, with two blackish, divergent humps in throat; side lobes partly hidden by the hood. HABITAT: dry, sandy places, heaths, woods; Spain to Greece; North Africa. March–May.

S. neglecta de Not. 309

Rather similar to *S. cordigera*, 10–30 cm. tall, with roughly heart-shaped lip but no spots on the leaves. Bracts shorter than flowers, usually flushed violet. Spike short, with 2–8 flowers 30–40 mm. long. Hood reddish violet, paler outside. Lip pale scarlet on margins, yellowish in centre, hairy; central lobe rather oval in outline, with parallel humps in throat, side lobes partly hidden by hood. HABITAT: dry places, garigue, always close to the sea; France, N. Italy, Corsica, Sardinia, possibly Sicily, Corfu. March–April.

S. orientalis *Nelson* is a similar plant but with larger lip which is typically more yellowish, sometimes reddish buff. HABITAT: S. E. Italy, S. Greece and the islands, Cyprus. March–April.

S. pseudocordigera *Moric.* (*S. vomeracea Briq., S. longipetala Poll.*) 310
Another stout, often tall species, 20–50 cm. high, with narrow, channelled, pointed leaves reflexing outwards. Flowers 4 to 10 or rarely more, 35 mm. long. Bracts markedly longer than flowers, of similar colouring. Hood clear pale red with deeper veins. Lip long, narrowly triangular (the Latin *vomeracea* means like a ploughshare). The side lobes are hidden, and there are two humps in the throat. The lip is brick red to reddish-brown and carries long hairs. HABITAT: marshes, damp meadows, scrub, on coasts and hills; circum-Medit. April–June.

S. lingua *L.* 311 TONGUE ORCHID
A slender plant, 10–25 cm. tall, rarely more. Leaves narrow, channelled, pointed, reflexing outwards. Stem with 2–6 flowers 30 mm. long, well spaced. Bracts about as long as flowers, reddish-violet. Hood projecting markedly forwards, with lip at right angles, held away from stem. Hood violet or flesh pink, often with green markings. Lip very variable in colour from very pale to deep rose pink, red or violet, rarely yellowish. Side lobes almost concealed; one large blackish hump in throat, sometimes grooved. HABITAT: open grassy or sandy places, scrub or light woodland; circum-Medit. March–June.

S. parviflora *Parl.* (*S. occultata Gay, S. laxiflora Chaub.*)
A smallish plant, usually 10–20 cm. tall, rarely more, with narrow, slightly waved leaves. Flowers 3–8, only 15–20 mm. long, with pointed bracts of the same length, which are red. Hood reddish-violet. Lip narrow, rusty red, hairy, pointing back to stem, the side lobes partly concealed, with parallel humps in throat. HABITAT: grassy, sandy or stony places, by coast; circum-Medit. April–June.

ACERAS

A. anthropophorum *(L.) Ait.* 306 MAN ORCHID
A rather slender plant, 10–40 cm. tall or rarely more, with crowded, keeled, glossy leaves, unmistakable with its long narrow spike of close-packed 15 mm. flowers, in which the yellow, greenish-yellow or maroon-tinged lip has two lateral and two slightly shorter terminal, strap-shaped divisions recalling the shape of a man. The French call the plant even more aptly the 'Hanged Man': his 'head' is composed of a greenish, rounded

242

hood. HABITAT: calcareous meadows, rocky hillsides, open scrub or woods; circum-Medit, dubiously in Palestine. April–July. A British native.

ANACAMPTIS

A. pyramidalis (L.) Rich.   307   PYRAMIDAL ORCHID
A slender plant 20–50 cm. tall, with long, narrow, keeled, pointed leaves. Flower head conical at first, expanding to an oblong shape, so densely packed with the 12 mm. flowers that the small bracts are not visible at flowering time. The lateral sepals are broad and out-spread; the other three segments form a loose hood; and the lip has three roughly equal lobes. There are two small erect 'guide-plates' in the centre of the flower which help insects to locate the small aperture of the long (12 mm.), thin, downpointing spur. Petals and sepals are rosy purple, lip bright pink. HABITAT: dry meadows or open ground, sometimes on sand dunes; circum-Medit. April-July. British native.

The *ssp. brachystachys Boiss.* (307), common in the E. Medit., is a smaller plant with rounded flower head and pale pink flowers. It flowers February–April.

SELECT BIBLIOGRAPHY

Volumes with illustrations are marked *.

AUTHORITATIVE FLORAS

Boissier, P. E., *Flora Orientalis, 5 vols.* (1867–88)
Baroni, P. E., *Guida Botanica d'Italia* (1955)*
Coste, H., *Flora Descriptive et Illustrée de la France* (1901–06)*
Davis, P. H. (Ed.), *Flora of Turkey and the East Aegean Islands* (1965–85)
Fiori, A., *Iconographia Florae Italicae* (1895–1904, 1933)*
Fiori, A., *Nuova Flora Analitica d'Italia* (1923–29)
Fournier, P., *Les Quatre Flores de la France* (1961)*
Halácsy, E. von, *Conspectus Florae Graecae* (1902–04)
Hayek, A., *Prodromus Florae Peninsulae Balcanicae* (1927–33)
Jahandiez, E., and Maire, R., *Catalogue des Plantes du Maroc* (1931–41)
Lazaro e Ibiza, B., *Compendio de la Flora Española* (1920)
Maire, R., *Flore de l'Afrique du Nord* (In production)*
Meikle, R. D., *Flora of Cyprus* (1977, 1985)
Nelson, E., *Monographie und Ikonographie der Gattüng Ophrys* (1962)*
Nelson, E., *Monographie und Ikonographie der Orchidaceen-Gattüng Dactylorhiza* (1976)*
Nelson, E., *Monographie und Ikonographie der Orchidaceen-Gattüng Serapias, Aceras, Loroglossum, Barlia* (1968)*
Quezel, F., and Santa, S., *Nouvelle Flore de l'Algérie* (1962)*
Rechinger, K. H., *Flora Aegaea* (1943)*
Tackholm, Vivi., *Student's Flora of Egypt* (1956)*
Willkomm, H. M., and Lange, J., *Prodromus Florae Hispanicae* (1861–80)
Zohary, M., and Feinbrun-Dothan, N., *Flora Palaestina* (1966–86)*

ECOLOGICAL STUDIES

Flahault, C., *La Distribution Geographique des Vegetaux dans la Région Mediterranéenne Française* (1937)
Harant, H., and Jarry, D., *Guide du Naturaliste dans le Midi de la France* (1961)*
Holmboe, J., *Studies on the Vegetation of Cyprus* (1914)*
Touring Club Italiano, *Conosci l'Italia, vol. II – La Flora* (1958)*
Turrill, W. B., *The Plant Life of the Balkan Peninsula* (1929)
Zohary, M., *Plant Life of Palestine* (1962)*

POPULAR ACCOUNTS AND SELECTIVE FLORAS

Atchley, S. C., *Wild Flowers of Attica* (1938)*
Brangham, A. N., *The Naturalist's Riviera* (1962)*
Davies, P. & J., and Huxley, A., *Wild Orchids of Britain and Europe* (1983)*
Huxley, A., and Taylor, W., *Flowers of Greece and the Aegean* (1977)*
Marret, L., *Les Fleurs de la Côte d'Azur* (1926)*
Moldenke, H. N., & A. L., *Plants of the Bible* (1952)*
Penzig, O., *Flore Coloriée de Poche du Littoral Mediterranéen de Gênes à Barcelone* (1902)*
Polunin, O., and Smythies, B. E., *Flowers of South-west Europe* (1973)*
Polunin, O., *Flowers of Greece and the Balkans* (1980)*
Theophrastus, *Enquiry into Plants*, trans. Sir Arthur Hort (1916)
Thompson, H. S., *Flowering Plants of the Riviera* (1914)*

INDEX

In this index to the descriptive text entries in capitals are to families and genera, and the remainder to species, except those in italics which are Latin synonyms or English names.

The bold italic numerals refer to the illustrations (numbers 1 to 311 in colour, between pages 22 and 25, and numbers 312 to 439 the line drawings on pages 27 to 50); the other numerals are references to the text pages.

253

INDEX OF POPULAR FOREIGN NAMES

GENUS	SPANISH	FRENCH	ITALIAN	GREEK
Acacia	Acacia, Mimosa	Mimosa	Gaggia	Akakía
Acanthus	Acanto	Acanthe	Acanto	
Agave	Pita	Agave	Agave	Athánatos
Anemone	Anémone	Anémone	Anemone	Agriopaparoúna, Anemóni
Arbutus	Madroño	Arbousier	Albatro, Corbezzolo	Koumariá, Lagomiliá
Asphodelus	Gamón	Asphodèle	Asfodelo	Spherdoúkla, Asphedeliá
Borago	Borraja	Bourrache	Borrana, Borragine	
Calendula	Maravilla, Caléndula, Flamenquilla	Souci	Cappuccina dei campi	
Calicotome	Retama espinosa	Cytise épineux	Ginestra spinosa	Aspálathos, Aspalathiá
Capparis	Alcaparra	Câprier	Cappero	Káppari
Carpobrotus	Flor de cuchillo	Mésembrianthème	Fico degli Ottentotti	
Ceratonia	Algarrobo	Caroubier	Carrubo	Xylokeratiá, Kharoupiá
Cercis	Arbol del amor	Arbre de Judée	Albero di Giuda, Siliquastro	Koutsoupiá, Koutsoukiá
Chamaerops	Palmito		Palma nana	
Chrysanthemum	Pajitos	Chrysanthème	Bambagella	Mandilída, Agriomantilída
Cichorium	Achicoria	Chicorée	Cicoria, Radicchio	Radíki
Cistus (various)	Jara, Jaguarzo	Ciste	Cisto	Kounoúklá, Ladaniá
Citrus limon	Limonero	Citronnier	Limone	Lemoniá
nobilis (C. deliciosa)	Mandarino	Mandarine	Mandarino	Mantariniá
sinensis	Naranjo	Oranger doux	Arancia	Portokaliá
Colutea	Espantalobos	Baguenaudier	Vescicaria	Phouska, Agriosinamikó
Crocus	Azafrán	Crocus	Zafferano, Croco	Crócos, Krináki
Cupressus	Ciprés	Cyprès	Cipresso	Kyparíssi
Cyclamen		Cyclamen	Ciclamino	Cyclámino, Cyclamiá
Cytisus	Piorno	Cytise	Maggio ciondolo	
Dracunculus	Dragontea, Serpentina	Serpentaire	Serpentaria, Dragontea	Drakontiá, Phidóchorto
Ecballium	Cohombrillo amargo, Pepinillo del diablo	Momordique, Giclet	Cocomero asinino, Schizzetti, Sputaveleno	Pikrangouriá, Agriangouriá
Erica	Brezo	Bruyère	Scopa	Ríki, Richiá
Eriobotrya	Níspero de Japón	Bibacier, Néflier du Japon	Nespolo del Giappone	Mousmouliá, Meskouliá

258

Ficus	Higuera	Figuier	Fico	Fíkos
Gladiolus	Gladiolo	Glaïeul	Gladiolo	Spathóhorto, Maïs
Helichrysum	Siempreviva	Immortelle	Elicriso, Ambrenti	Amáranto
Hyoscyamus	Beleño	Jusquiame	Giusquiamo	Discýamo, Gérontas
Iris	Lirio	Iris	Giaggiolo	Agriókrinos, Krinákia, Vourlítis, Vroúla
Juniperus	Enebro, Sabina	Genévrier	Ginepro	Kédro, Agriokyparíssi
Laurus	Laurel	Laurier	Alloro, Lauro	Dáphni, Vaïá
Lavandula	Lavanda	Lavande	Lavanda, Spico	Levánta, Livanáki
Limonium	Acelga silvestre	Lavande-de-mer	Statice	
Linum	Lino	Lin	Lino	Linári, Agriolínaro
Lonicera	Madreselva	Chèvrefeuille	Periclimeno, Madreselva, Caprifoglio	Hagióklima, Agrióklimon
Lupinus	Altramuz	Lupin	Lupino	Loúpino, Loupinári
Mandragora	Mandrágora, Berengenilla	Mandragore	Mandragora	Mandragóras, Mandragoúda
Matthiola	Alhelí	Violier	Violacciocca	Violétta
Muscari	Jacinto de penacho	Muscari	Giacinto	
Myrtus	Mirto, Murta	Myrte	Mirto, Mortella	Myrtiá, Smirtiá
Nerium	Adelfa, Baladre	Laurier rose	Oleandro, Leandro	Pikrodáphne, Rhododáphne
Olea	Acebuche, Olivo	Olivier	Olivo	Eliá
Ophrys	Abejera	Ophrys	Fior ragno	
Opuntia	Chumbera	Figuier de Barbarie	Fico d'India	Fragosykiá
Orchis		Orchis		Salépi, Sernikovótano
Orobanche		Orobanche	Succiamele, Erba lupa, Fiamma	Liýkos
Paliurus	Espina de Cristo	Épine du Christ	Paliuro, Marruca	Palioúri
Pancratium	Nardo marino	Lis-mathiole	Pancrazio, Narciso marina	Krinos tis thálassas
Papaver	Amapola, Ababol	Pavot, Coquelicot	Papavero, Rosolaccio	Paparoúna, Koutsounáda
Phlomis	Aguavientos	Herbe-au-vent	Flomide	Aspháka, Alisphakiá
Pinus	Pino	Pin	Pino	Pévko
Pistacia				
lentiscus	Lentisco	Lentisque	Lentisco	Skinos, Skinári
terebinthus	Cornicabra	Térébinthe	Terebinto	Kokkorevithiá

Platanus	Plátano	Platane	Platano	Plátanos
Prunus dulcis	Almendro	Amandier	Mandorlo	Amygdaliá
Punica	Granado	Grenadier	Melograno, Granato	Roïdiá, Rodiá
Quercus				
coccifera	Coscoja	Chêne-kermès	Querco spinosa	Prinos, Prinári, Pournári
ilex	Encina, Carrasca	Yeuse	Leccio	
suber	Alcornoque	Chêne-liège	Sughera	
Rosmarinus	Romero	Romarin	Rosmarino	
Ruta chalepensis	Ruda	Rue	Ruta	Apíganos
Salvia	Salvia	Sauge	Salvia	Faskomiliá, Alisphakia Stivída, Apháná, Astívi
Sarcopoterium			Spina porci	
Schinus	Pimentero falso	Faux poivrier	Pepe, Falsopepe	Skólympros, Asprágatho
Scolymus	Tagarnina	Épine-jaune	Cardo scolimo, Scardiccione	
Serapias		Sérapias	Limodoro	Glossári
Silybum	Cardo lechero, C. de María	Chardon Marie	Cardo mariano	
Smilax	Zarzaparilla	Liseron épineux	Smilace, Salsapariglia nostrale	Arkoudóvatos, Arkóvatos
Spartium	Retama de olor	Genêt d'espagne	Ginestra	Spárto
Styrax		Aliboufier	Storace	Astýrakas, Stourakiá, Lagomiliá
Tamarix	Taray, Atarce	Tamarin	Tamerice	Lagóhorto
Thymelaea	Bufalaga	Passerine	Barbosa, Spazzaforno	Thymári
Thymus	Tomillo	Thym	Timo	Pigounítis, Lagóhorto
Tragopogon	Barba cabruna	Barbe-de-bouc	Barba di becco	Lales
Tulipa	Tulipán	Tulipe	Tulipano	Skyllokrómmýda
Urginea	Cebolla albarrana	Scille maritime, Oignon marin	Scilla	
Verbascum		Molène, Bouillon blanc, Bonhomme	Verbasco	Phlómos
Viburnum tinus	Durillo	Laurier-tin	Viburno, Lauro Tino	
Vinca		Pervenche	Pervinca	Agrioliza
Vitex	Agnocasto	Gattilier	Vitice, Agno casto	Lygariá, Alygariá
Vitis	Vid, Parra	Vigne	Vite	Ampéli
Zizyphus	Azufaifo	Jujubier	Giuggiolo, Zizzolo	Tzitziá